P9-BIV-108

California

Mapping the Golden State through History

Rare and Unusual Maps from the Library of Congress

Vincent Virga

and Ray Jones

Guilford, Connecticut

For Fran Siegel, who helped me see the golden light of California for what it really is—a promise.
—Ray Jones

Text design: Sheryl P. Kober
Layout: Casey Shain
Project editor: John Burbidge

Library of Congress Cataloging-in-Publication Data
Virga, Vincent.
California : mapping the golden state through history : rare and unusual maps from the Library of Congress / Vincent Virga and Ray Jones.
p. cm.
Includes bibliographical references.
ISBN 978-0-7627-4530-2
1. California—Historical geography—Maps. 2. California—History—Maps. 3. Early maps—California—Facsimiles. 4. California—Maps, Manuscript—Facsimiles. I. Jones, Ray. II. Library of Congress. Geography and Map Division. III. Morris Book Publishing (Firm) IV. Title.
G1526.S1V5 2009
912.794--dc22
2009018279

Printed in China

10 9 8 7 6 5 4 3 2 1

Contents

Foreword

BY VINCENT VIRGA

MYTHS AND LEGENDS PLAYED A HUGE PART IN the world's imagination in 1562 when the name "California" made its debut on a map by Diego Gutiérrez. It was as famous a place in Europe as the subject of his map, America, and it was no longer an island. At a time when scientific thought was resurfacing from the crushing weight of Christian texts that had dominated the medieval mind, fantastical travel memoirs, such as Marco Polo's book, written circa 1298, and John Mandeville's book, written in 1357, still held sway. To their enshrined myths and legends, a romance novel from 1510 by Garci Rodríguez de Montalvo added California: an island inhabited by black Amazons with pots of gold!

Our author Ray Jones correctly contends, "California is not just a place; it is a state of mind." How can it be otherwise when even the derivation of its name is up for grabs. It was also far, far away from the mapmakers of Europe, but as maps represent dreams, ideas, actions, and records of human endeavors, the communal imagination assisted our mapmakers in discovering geographical reality by motivating adventurers to journey into the vast unknown, to the ends of the earth.

Like maps, California *is* dreams, ideas, actions, a record of human endeavors. It gave us gold rushes, earthquakes, and gods and goddesses on a silver screen. Its place in the American imagination is immense, in part because it harbored the dream makers who formulated the American Dream—our movie makers.

Living on planet Earth has always raised certain questions from time to time for those of us so inclined. Of course, the most obvious one is, where am I? Well, as Virginia Woolf sagely noted in her diary, writing things down makes them more real, and this may have been a motivating factor for the Stone Age artists who invented the language of signs on the walls of their caves in southern France and northern Spain between 37,000 and 11,000 years ago. Picasso reportedly said, "They've invented everything," which includes the very concept of an image.

A map is an image. It makes the world more real for us and uses signs to create an essential sense of place in our imagination. Cartographic imaginings not only locate us on this earth but also help us invent our personal and social identities, since maps embody our social order. As did the movies, maps helped create our national identity, and this encyclopedic series of books aims to make manifest the changing social order that invented the United States, which is why the series embraces all fifty states.

Each is a precious link in the chain of events that is the history of our "great experiment," the first and enduring federal government ingeniously deriving its just powers—as John Adams proposed—from the consent of the governed. Each state has a physical presence that holds a unique place in any representation of our republic in maps. To see each one rise from the body of the continent conjures Tom Paine's excitement over the resourcefulness of our Enlightenment philosopher-founders: "We are brought at once to the point of seeing government begin, as if we had lived in the beginning of time." Just as the creators systemized not only laws but also rights in our Constitution, so our maps show how their collective memory inspired the body politic to organize, codify, and classify all of nature to do their bidding, with passionate preferences state by state. For they knew, as did Alexander Pope:

> *All are but parts of one*
> *stupendous Whole*
> *Whose body Nature is, and*
> *God the soul.*

And aided by the way maps under interrogation often evoke both time and space, we editors and historians have linked the reflective historical overviews of our nation's genesis to the seduction of place embedded in the art and science of cartography.

J. P. Harley posits, "The history of the map is inextricably linked to the rise of the nation-state in the modern world." The American bald eagle has been the U.S. emblem since 1782, after the Continental Congress appointed a committee in 1776 to devise an official seal for our country. The story of our own national geographical writing begins in the same period but has its roots centuries earlier, appropriately, in a flock of birds.

On October 9, 1492, after sailing westward for four weeks in an incomprehensibly vast and unknown sea during North America's migration month, an anxious Christopher Columbus spotted an unidentified flock of migrating birds flying south and signifying land—*"Tierra! Tierra!"* Changing course to align his ships with this overhead harbinger of salvation, he avoided being drawn into the northern-flowing Gulf Stream, which was waiting to be charted by Ben Franklin around the time our eagle became America as art. And so, on October 11, Columbus encountered the salubrious southern end of San Salvador. Instead of coming ashore somewhere in the future New England, he came up the lee of the island's west coast to an easy and safe anchorage.

Lacking maps of the beachfront property before his eyes, he assumed himself to be in

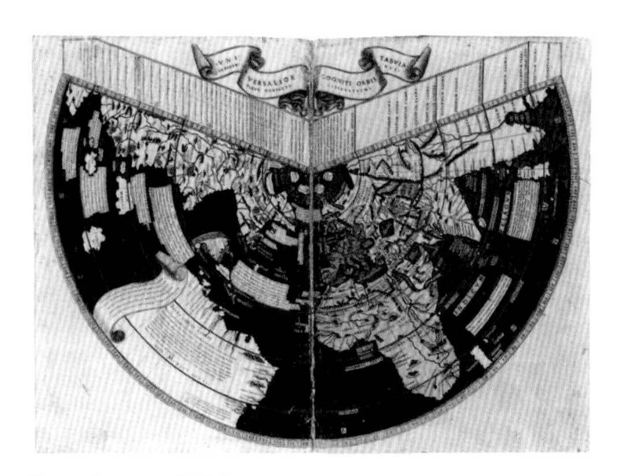

Ruysch map, 1507

Asia, because in his imagination there were only three parts to the known world: Europe, Asia, and Africa. To the day he died, Columbus doubted he had come upon a fourth part, even though Europeans had already begun appropriating, through the agency of maps, what to them was a New World. Perhaps the greatest visual statement of the general confusion that rocked the Old World as word spread of Columbus's interrupted journey to Asia is the Ruysch map of 1507. Here we see our nascent home inserted into the template created in the second century by Ptolemy, a mathematician, astrologer, and geographer of the Greco-Roman known world, the *oikoumene.*

This map changed my life. It opened my eyes to the power of a true cultural landscape. It taught me that I must always *look* at what I *see* on a map, focusing my attention on why the map was made, not who made it, or when or where it was made, but *why.* The Ruysch map was made to circulate the current news. It is a quiet meditative moment in a very busy, noisy time. It is life on the cusp of a new order. And the new order is what Henry Steele Commager christened the "utopian romance" that is America. No longer were maps merely mirrors of nature for me. No longer were the old ones "incorrect" and ignorant of the "truth." The Ruysch map is reality circa 1507! It is a time machine. It makes the invisible past visible. Blessedly free of impossible abstractions and idealized virtues, it is undeniably my sort of primary historical document.

The same year, 1507, the Waldseemüller map appeared. It is yet another reality and one very close to the one we hold dear. There we Americans are named for the first time. And there we sit, an independent continent with oceans on both sides of us, six years *before* Balboa supposedly discovered "the other sea." There are few maps as mysterious for cartographic scholars as Waldseemüller's masterpiece. Where did all that news come from? For our purposes it is sufficient to say to the world's visual imagination, "Welcome to us Americans in all our cartographic splendor!"

Throughout my academic life, maps were never offered to me as primary documents. When I became a picture editor, I learned to my amazement that most book editors function solely as "word people." Along with historians and academics, they make their livelihood working with words and ideas. The fact of my being an "author" makes me a

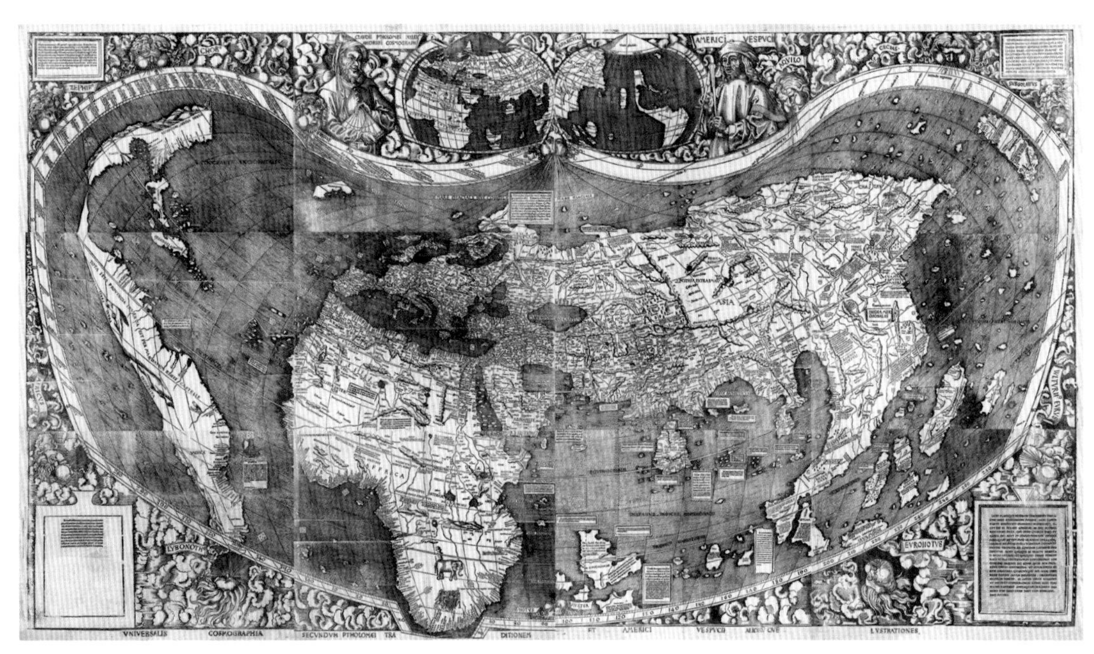

Waldseemüller map, 1507

word person, too, of course. But I store information visually, as does a map. So I am both a word person and a picture person.

The very title of this volume, *California: Mapping the Golden State through History,* makes it clear that this series has a specific agenda, as does each map. It aims to thrust us all into a new intimacy with the American experience by detailing the creative process of our nation in motion through time and space via word *and* image. It grows from the relatively recent shift in consciousness about the physical, mental, and spiritual relevance of maps in our understanding of our lives on earth. Just as each state is an integral part of the larger United States, "Where are we?" is a piece of the larger puzzle called "Who are we?"

The Library of Congress was founded in 1800 with 740 volumes and three maps. It has grown into the world's largest library and is known as "America's Memory." There are nearly 5 million maps in the Geography and Map Division. We have linked our series with that great collection in the hopes that its astonishing breadth will inspire us in our efforts to strike Lincoln's "mystic chords of memory" and create living history.

On January 25, 1786, Thomas Jefferson wrote, "Our confederacy must be viewed as the nest from which all America, North and South is to be peopled." This is a man who could not live without books. This is a man who drew maps. This is a politician who, in spite of his abhorrence of slavery and his respect for Native Americans, took pragmatic rather than principled positions when confronted by both "issues." Nonetheless, his bold vision of an expanded American universe informs our current enterprise. There is no denying that the story of the United States has a dark side. What makes the American narrative unique is the ability we have displayed time and again to remedy our mistakes and adjust to changing circumstances.

For Jefferson, whose library was the basis for the current Library of Congress after the British burned the first one during the War of 1812, and for his contemporaries, the doctrine of progress was a keystone of the Enlightenment.

The maps in our books are reports on America, and all of their political programs are manifestations of progress. Our starting fresh, free of Old-World hierarchies, class attitudes, and the errors of tradition, is wedded to our geographical isolation with its immunity from the endless internal European wars devastating humanity, which justifies Jefferson's confessing, "I like the dreams of the future better than the history of the past." But as the historian Michael Kammen explains, "For much of our history we have been present-minded; yet a usable past has been needed to give shape and substance to national identity." Historical maps keep the past warm with life. They encourage critical inquiry, curiosity, and qualms.

For me this series of books celebrating each of our states is not about the delineation of property rights. It is a depiction of the pursuit of happiness, which is listed as one of our natural rights in the 1776 Declaration of Independence. Jefferson also believed that "the earth belongs always to the living generation." I believe these books depict what each succeeding generation in its pursuit of happiness accomplished on this portion of the earth known as the United States. If America is a matter of an idea, then maps are an image of that idea.

I also fervently believe these books will show the states linked in the same way Lincoln saw the statement that all men are created equal: as "the electric cord in that Declaration that links the hearts of patriotic and liberty-loving men together, that will link those patriotic hearts as long as the love of freedom exists in the mind of men throughout the world."

VINCENT VIRGA
WASHINGTON, D.C.
INAUGURATION DAY, 2009

Introduction

California is not just a place; it is a state of mind, and this has always been the case. Europeans dreamed of California long before their explorers ever set eyes upon it. In the vivid imaginations of the renaissance-era Spanish, French, and Italians, California was a misty island where easy riches and mysterious delights awaited them—if they only knew where to find it.

In 1510 myth spinner Garci Rodríguez de Montalvo enraptured hordes of readers with tales of California in his romance novel entitled *Las Sergas de Esplandián* (The Adventures of Esplandián). Montalvo described California as an island inhabited by a race of powerful warrior women. Concerning these women and their home, he wrote:

> *On the right hand of the Indies is an island called California, very near to the Terrestrial Paradise and inhabited by black women without a single man among them and living in the manner of the Amazons. They are robust of body, strong and passionate of heart, and of great valor. Their island is one of the most rugged in the world with bold rocks and crags. Their arms are all of gold as are the harnesses of the wild beasts, which after taming, they ride. In all the island there is no other metal.*

Some may not have taken such fanciful descriptions seriously, but many did. More than a few of the young Spaniards who boarded leaky ships and set sail for the Caribbean after Columbus discovered America did so in hopes of finding California. Most came first to Cuba, which was, of course, a large island, but it certainly was not California. For one thing, it had no gold.

In 1519, less than thirty years after Columbus discovered America, a few hundred Spanish adventurers led by Hernán Cortés set out from Cuba and landed on

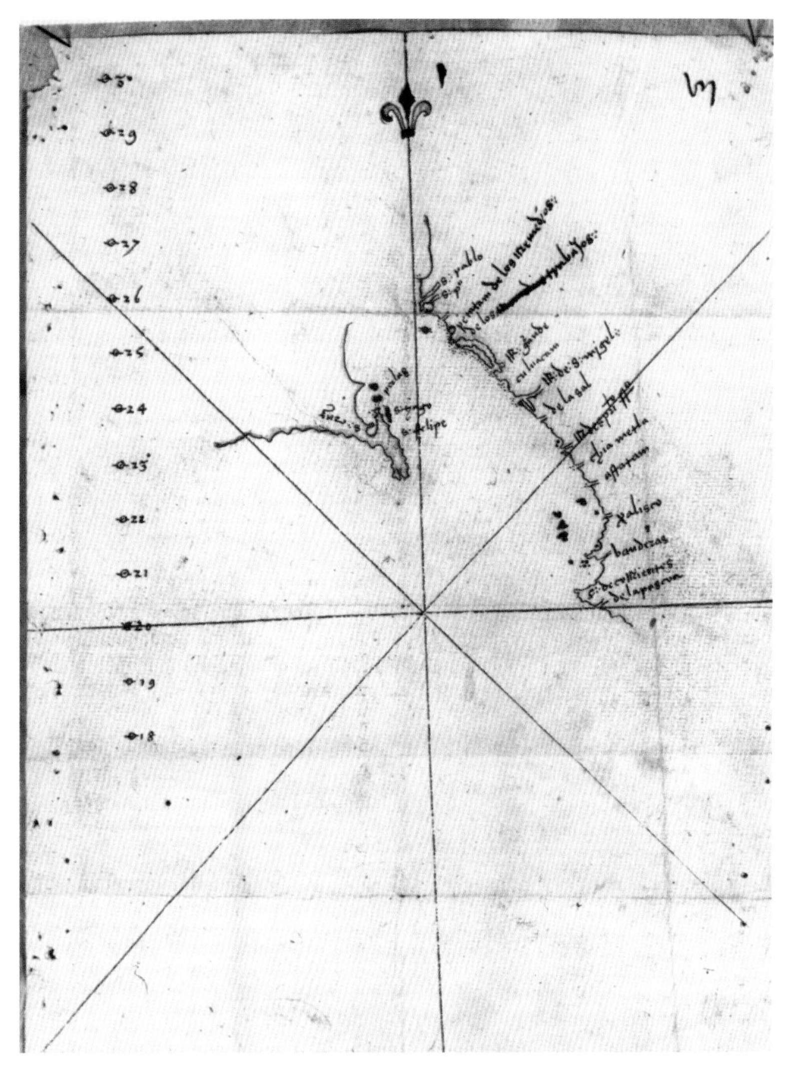

Mexico (West Coast) (1535).

Having conquered the Aztecs, Hernán Cortés sent explorers to the west coast of Mexico, where they built small ships and ventured into the Gulf of California. They soon discovered the Baja California peninsula, which Cortés may have mistaken for the mythic "island" of California. The conquistador was likely disappointed to find no gold or Amazon women when he visited the peninsula in 1535. Produced that same year, this crudely drawn map shows key landfalls along the gulf's eastern shore as well as the southern tip of the peninsula. A conspicuous detail is La Paz Bay, where Cortés's ships sought shelter. Note that the map is open ended on the north, suggesting that Cortés and his men believed the gulf might extend northwestward all the way to the Pacific.

the shores of what is today Mexico. Their primary objective was to search for gold, if not to find California itself. Pushing rapidly inland, they defeated the mighty Aztecs and founded the first major European colony in North America.

Cortés must have been a fan of Montalvo's book, for he soon dispatched maritime expeditions to search for the novelist's mythic island. Cortés's ships explored the 1,000-mile-long Gulf of California and the shores of the extensive peninsula known today as Baja California. Other ships sailed across the Pacific as far as the Philippines. However, none of these mariners discovered any gold-laden islands populated by strange beasts and muscular black Amazons. Even so, belief in this fabled place persisted, and for centuries some of the less-well-informed European mapmakers continued to show California as an island.

Eventually, frustrated explorers must have concluded that Montalvo's island paradise did not exist. They came to regard it as a fictional invention that had sprung from the overactive imagination of a romanticist. Almost certainly this was the case, but given time, fiction and reality can produce some very bizarre coincidences. After all, history and myth are parallel currents running through the same river. As it turned out, the real California, when it was finally discovered and then settled beginning in the eighteenth century, would bear curious similarities to the mythical one.

The Real Island of California

Although not surrounded by water, California is in every other sense of the word an island. Instead of ocean waves, the arid sea of land we know today as the Desert Southwest separates California's substantial landmass from the rest of the continent.

California is a cultural island as well, and this has been true for thousands of years. Before settlers of European heritage came here and devastated them with guns and diseases, California's native peoples lived mostly in small communities. There was little trade, and different groups maintained a strict separation, not just from tribes outside California but from one another as well. Often neighboring groups did not even speak the same language or worship the same spirits. People were forbidden to speak of the dead, and this precluded keeping any sort of historical record, written or otherwise. For these early Californians myth *was* history.

And as we have seen, the earliest maps of California were based on dreams rather than facts. They were drawn before the southwestern corner of the North American continent came to be called California and long before mapmakers had learned, much to their surprise, that there was no large island by that name. However, the charts and descriptions brought back from the New World by early explorers began to fill the gaps in their knowledge, and mapmakers created increasingly accurate depictions of the place we now call California. As a result, the mythical island of California slowly sank into the Pacific, but other totally inaccurate geographic notions proved even longer lasting. One of these was the so-called Strait of Anian—known to the British as the Northwest Passage.

The Strait of Anian was said to link the Atlantic and Pacific oceans. From the time of Columbus onward, Europeans searched for this nonexistent sea route, which they hoped would open the door to East Asia with its silk, spices, jewels, and other riches. In 1542 Juan Rodríguez Cabrillo set out from western Mexico in search of the strait. For months

Cabríllo sailed slowly northward along the coasts of Baja California and the modern-day state of California, exploring dozens of inlets and bays, but none of them promised to carry him to the Atlantic. Nor did he manage to circumnavigate what some still stubbornly believed was the island paradise of California. Cabríllo's expedition proved quite the opposite—that California was not an island but rather a bold continental coast fraught with dangers for mariners and uninviting to settlers.

For the next two hundred years, California attracted only a handful of Europeans. One was English privateer Francis Drake, who brought his ship, the *Golden Hind,* here in 1579 during his epic voyage around the world. In need of food and water, Drake and his men went ashore near Point Reyes, a rugged peninsula located about two dozen miles northwest of modern-day San Francisco. There they were greeted by Miwok Indians, who wailed and scratched their cheeks as if mourning the sudden appearance of these strange pale men from the sea. Apparently, the Miwoks thought the Englishmen were the souls of long-dead relatives.

Spanish seamen occasionally made unintended visits. The captains of galleons returning to Mexico from trading expeditions in the Far East sometimes made navigational errors that caused them to run unexpectedly onto the rock-strewn California coast—nearly always with disastrous results. In 1595 galleon captain Sebastián Rodríguez Cermeño explored the coast, looking for protected coves and bays to shelter lost or stricken Spanish vessels. In 1602 Sebastián Vizcaíno led a similar expedition. Cermeño and Vizcaíno located a few temporary safe harbors but made no effort to establish a permanent Spanish presence in California.

Nearly two centuries would pass before the Spanish made anything more than a halfhearted attempt to settle California. In time, however, pioneers would come, initially a mere trickle of them and then a torrent. First came the Spanish padres, then ranchers, then homesteaders from the East, then miners, and finally dream seekers from all over the world. As California dreamers had from the first, many of them came looking for gold, and their thirst for the bright yellow metal gave rise to one of the California story's greatest ironies. While not apparent to its earliest settlers, California did have gold, more of it, in fact, than Montalvo and his readers might ever have imagined.

Of course, not all California dreams have been golden, for they have taken on every color of the rainbow. People have come to California seeking riches derived from the black gold of oil; from the redwood of the sequoia; from brown chocolate and tan almonds; from yellow lemons and green limes; and, of course, from the silver screen.

For many, California itself is a dream. Every day in Chicago, Cleveland, Boston,

Dallas, Tokyo, New Delhi, Melbourne, Moscow, and Reykjavik, there are always at least a few who are tired of the local weather and of their lot in life and are giving serious thought to packing up and moving to California. They figure it's a place where you can finally let go and start being yourself. It's a place where the promises of life are kept. Does such a place truly exist? They are sure it does, for they have seen it on television and at the movies—and they have seen maps of it.

The Language of Maps

Maps are a peculiar human invention. The first were created thousands of years ago to give travelers some idea of what to expect along the coasts they were about to sail or in the hinterlands they were about to trek. Maps have always served these purposes, but they have accomplished many other things as well. They have given names to places that had never before had a fixed identity. They have also given whole populations of people an identity and given them cause for both hope and despair. They have raised some nations up and helped to destroy others, and while doing it have added immeasurably to the historical record.

One of the most intriguing things about maps is that they are so purely symbolic. A squiggly line on a page may represent a turning, twisting road, but it might just as easily be meant to suggest a winding river. And no map, not even the most painstakingly assembled twenty-first-century one, can ever be 100 percent accurate, in part because a map is already out of date the moment it is published. A map can only offer a snapshot, a suggestion of what things were like at a certain point in time. In fact, most maps are not intended to accurately describe spatial relationships. They are intended to make a point. A map of Los Angeles may be intended to convince tourists that the City of Angels is a wonderful place to visit, or to remind the public that the threat of wildfire should be taken very seriously.

The colorful and fascinating maps you'll see in *California: Mapping the Golden State through History* were selected by the Library of Congress and are handsomely reproduced on these pages for yet another reason. Their purpose here is to tell the dramatic story of California in an utterly unique and particularly revealing way. They are not meant to guide you through cities, along roads, or over mountain passes. Indeed, some of the roads and features depicted on these maps no longer exist or have been altered beyond all recognition. Rather it is hoped that these maps will help you negotiate the highways and byways of time. They are cartographic symbols, wavy lines intended to show you how California and Californians got to where they are today.

AMERICAE SIVE QVARTAE ORBIS PARTIS NOVA ET EXACTISSIMA
VICTORIA
SEPTENTRIO
PARTE DI GRONLANDA
PICNEMAY REG
TIERRA DE LABRADOR
CANADA PROV
TIERRA DE BACCALAOS
TIERRA FRANCISCA
TIERRA DI NORIMBERGA
AVACAL
APALCHEN
MOCOSA
VLTERIVS HEC TERRA INCOGNITA PERMANSIT
CALICVAZ
OTAGIL
CAPAS CHI
TIERRA FLORIDA
Monte Suala
LA NVEVA GALITIA
COSSA
TOPIRA
XALISCO
GOLFO DE LA NVEVA ESPAÑA
CV BA
NVEVA ESPAÑA
CASTILLA DEL ORO
BENEZVELA REG
PARIA
AEQVINOCTIALIS
NOVA ANDALVSIA
CARTAGENA
MAR DEL SVR
REGIO DEL PERV

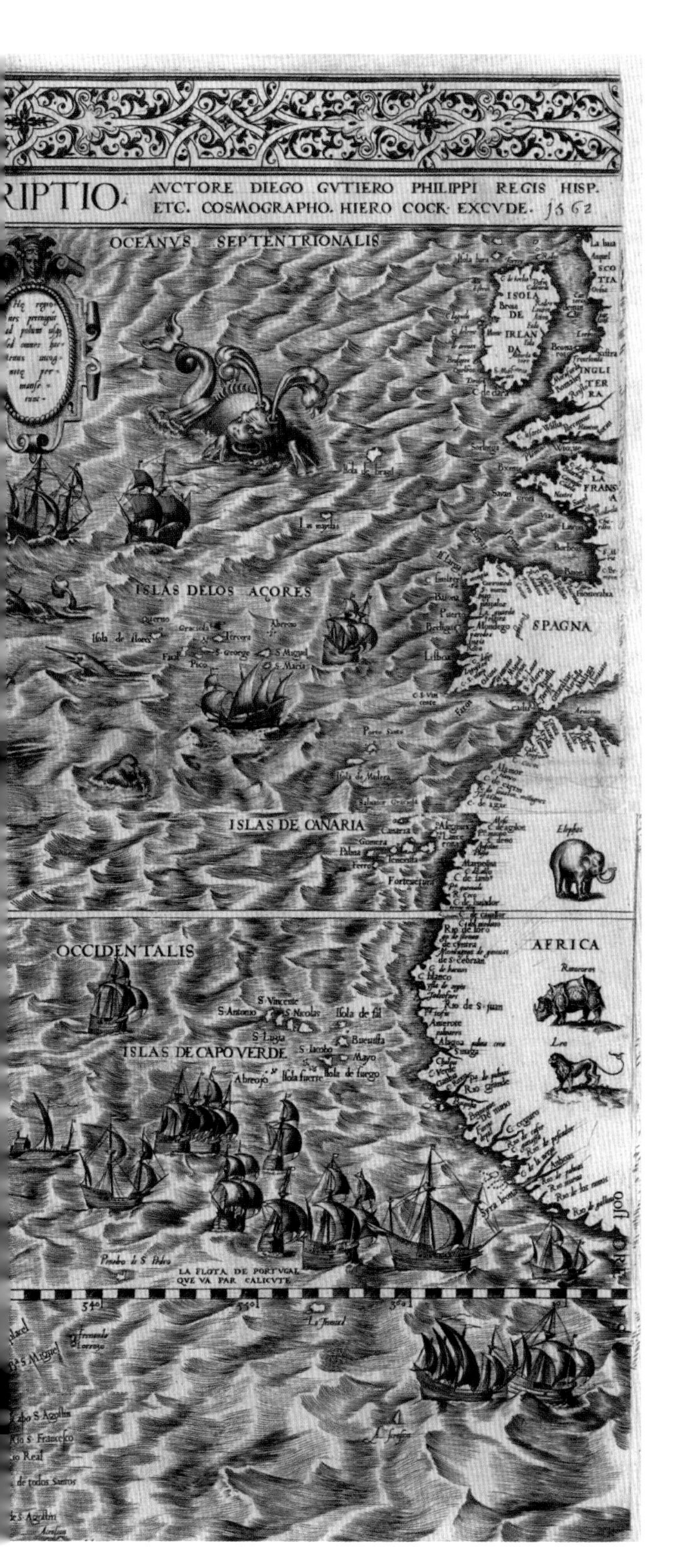

Americae sive Quartae Orbis Partis Nova et Exactissima Descriptio (1562).

By the mid-1500s Spanish explorers had provided enough information to give mapmakers a rough vision of North and South America. A chart and instrument maker who had served as a pilot on ships, Diego Gutiérrez produced this ornate view, which was published in 1562, several years after his death. Details such as the Amazon River, the Florida and Yucatan peninsulas, and the mountainous interior of North America show that Gutiérrez had a reasonably accurate grasp of the topography. However, the map also depicts sea monsters and other mythic marvels. This is the first map to include a mention of California (see far left, middle of inset), a place where Europeans expected to find abundant gold, wondrous beasts, and warrior women.

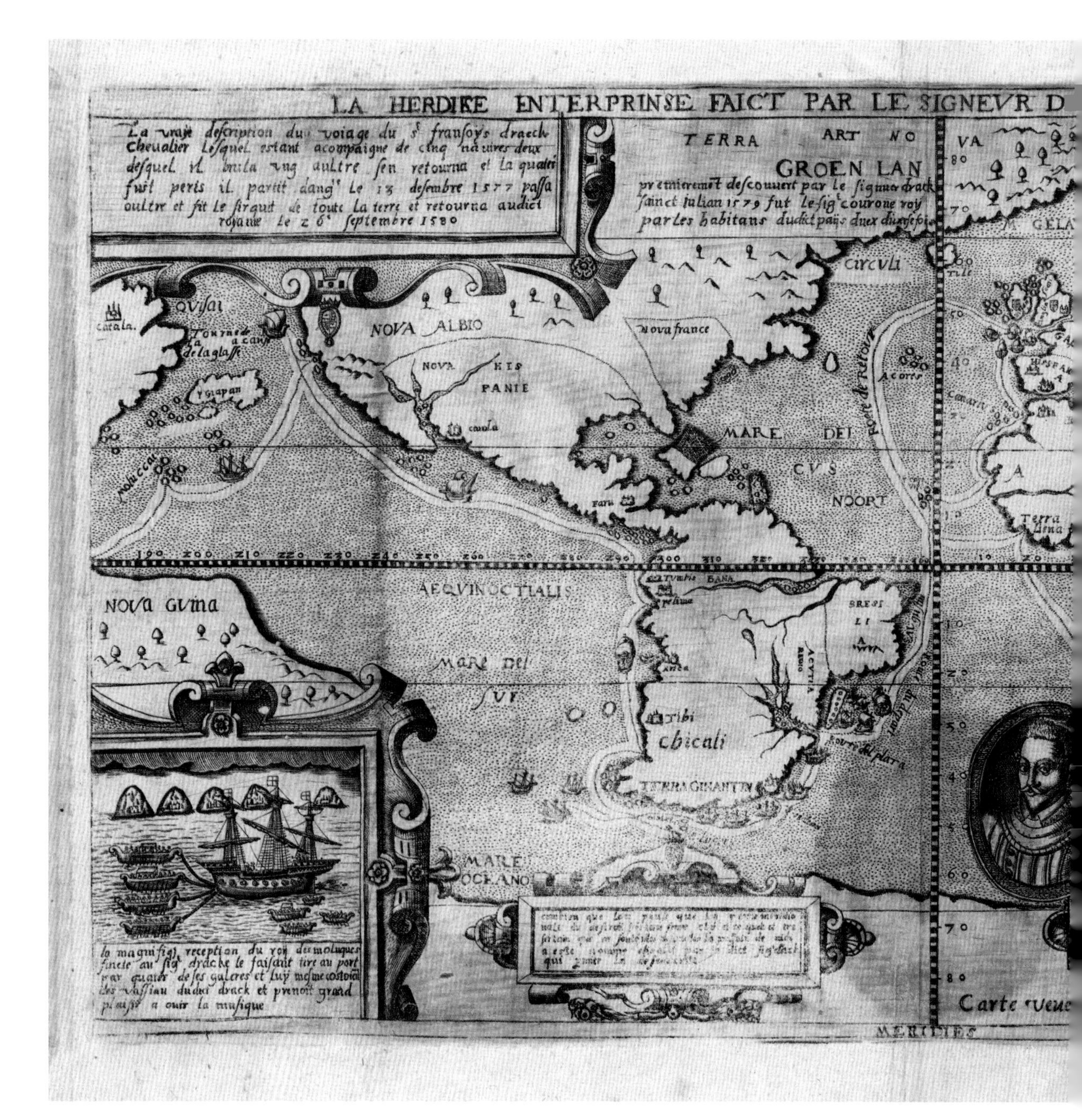
LA HERDIKE ENTERPRINSE FAICT PAR LE SIGNEVR D
TERRA ART NO
GROEN LAN
QVISAI
catala
NOVA ALBIO
nova france
NOVA KIS PANIE
MARE DEL
NOORT
AEQVINOCTIALIS
NOVA GVINA
MARE DEL SVR
chicali
TERRA GIGANTVM
MARE OCEANO
Carte veue

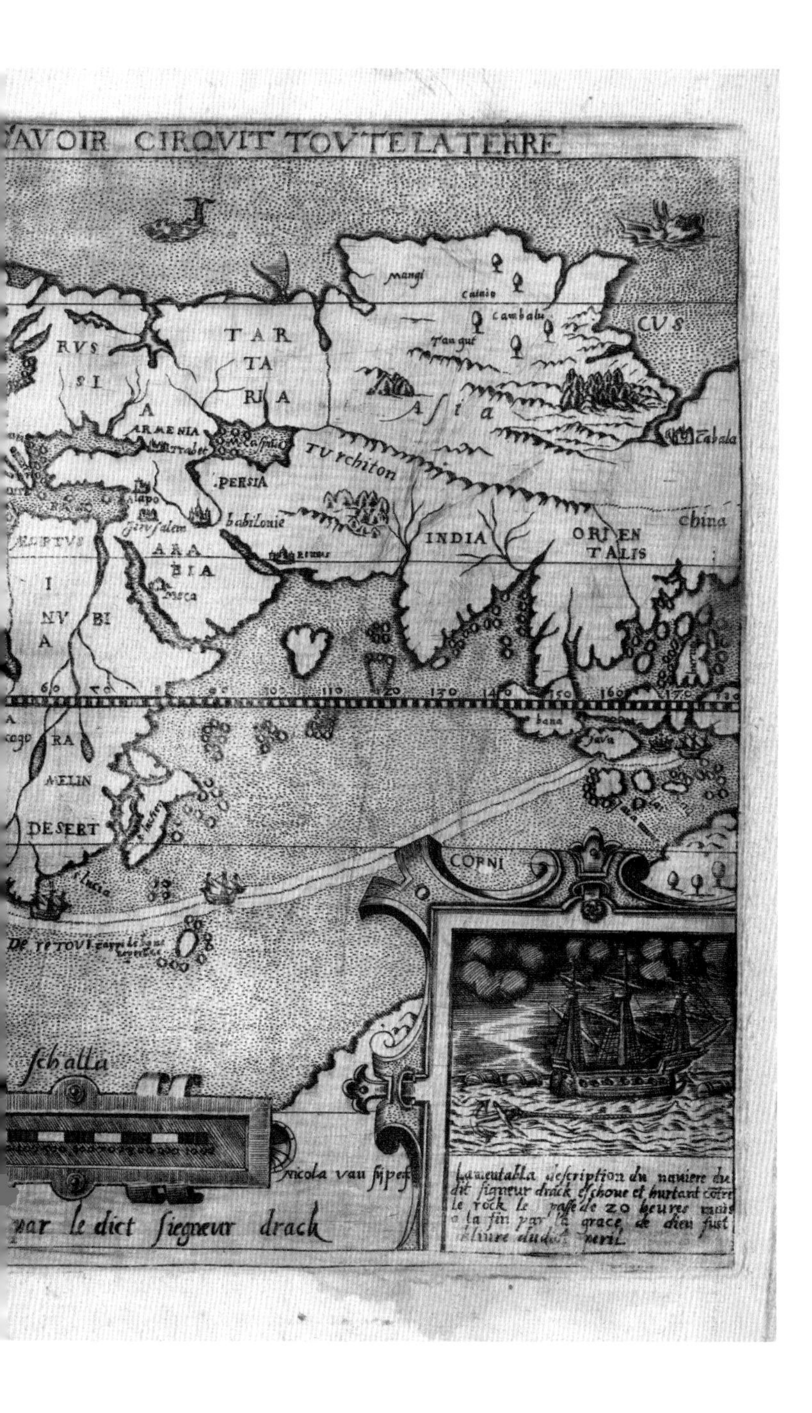

La herdike enterprinse faict par le Signeur Draeck D'Avoir cirquit toute la Terre / Nicola van Sype (1581). Dutch craftsman Nicola van Sype made this map in 1581 to commemorate Sir Francis Drake's voyage around the world, completed just two years earlier. Details are likely drawn from a map made during the voyage by Drake himself and presented to Queen Elizabeth I upon his return. The engraving shows that Drake brought his ship, the *Golden Hind,* to the west coast of North America, where he landed, most likely on or near Point Reyes, about 25 miles northwest of modern-day San Francisco. The Drake landing places him among the first Europeans to have set foot in California.

Map of California shown as an island, Joan Vinckeboons (1650).

Attributed to prolific Dutch mapmaker Joan Vinckeboons, this 1650 chart shows the "island" of California. As early as 1542 the expedition of Juan Rodríguez Cabrillo had proven California to be part of the North American landmass. Even so, some European cartographers continued to depict California as an island well into the eighteenth century. The distinctive shape of Vinckeboons's island and of the strait separating it from the mainland strongly suggests that the island myth was rooted, in part, in confusion over the true nature of the spindly and rugged peninsula known today as Baja California. The inset above details some of the islands off the coast of Baja.

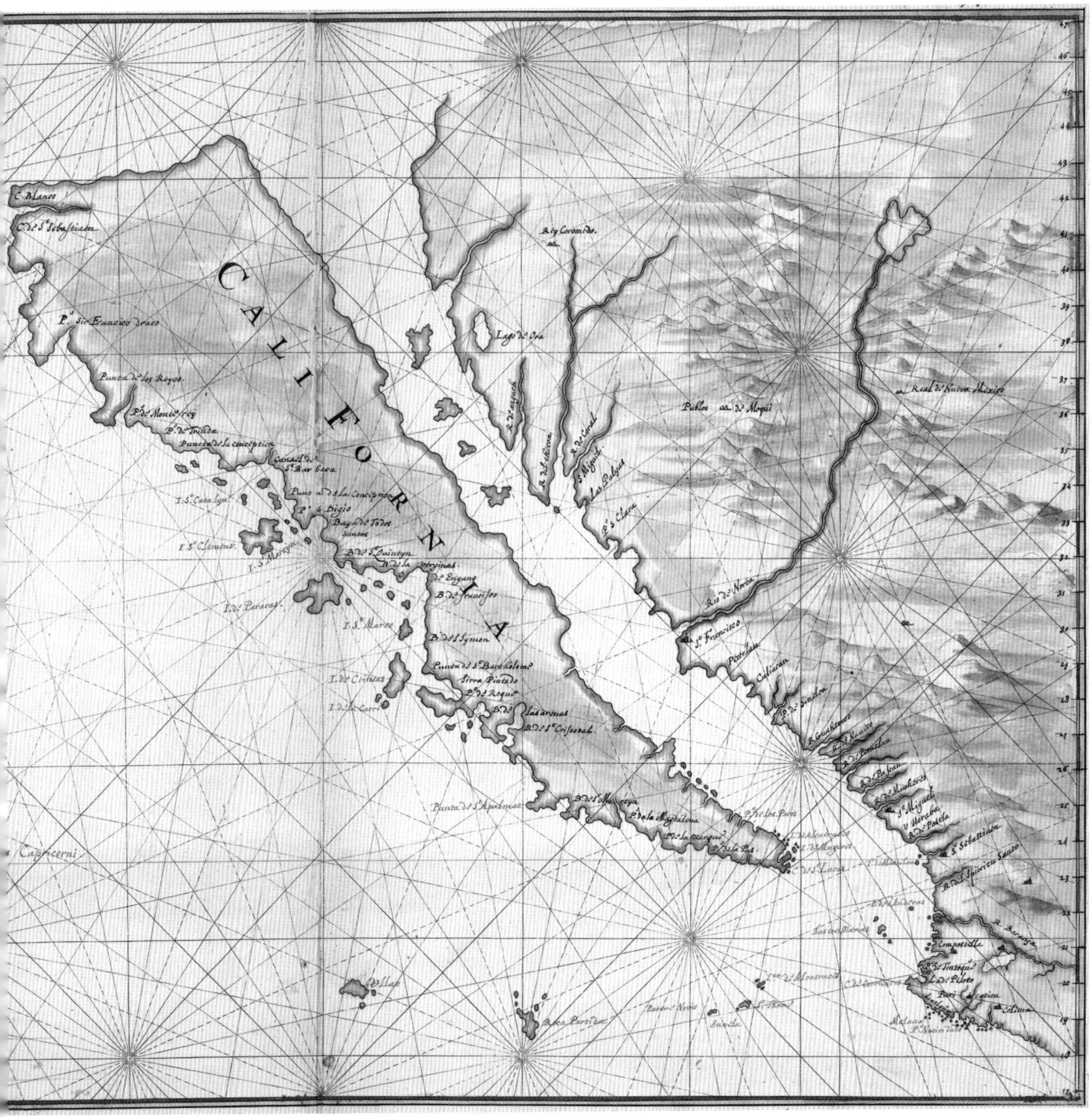
C. Blanco
CALIFORNIA
Lago de Ora
Rio de Norte
Capricorni
Compostella
Colima

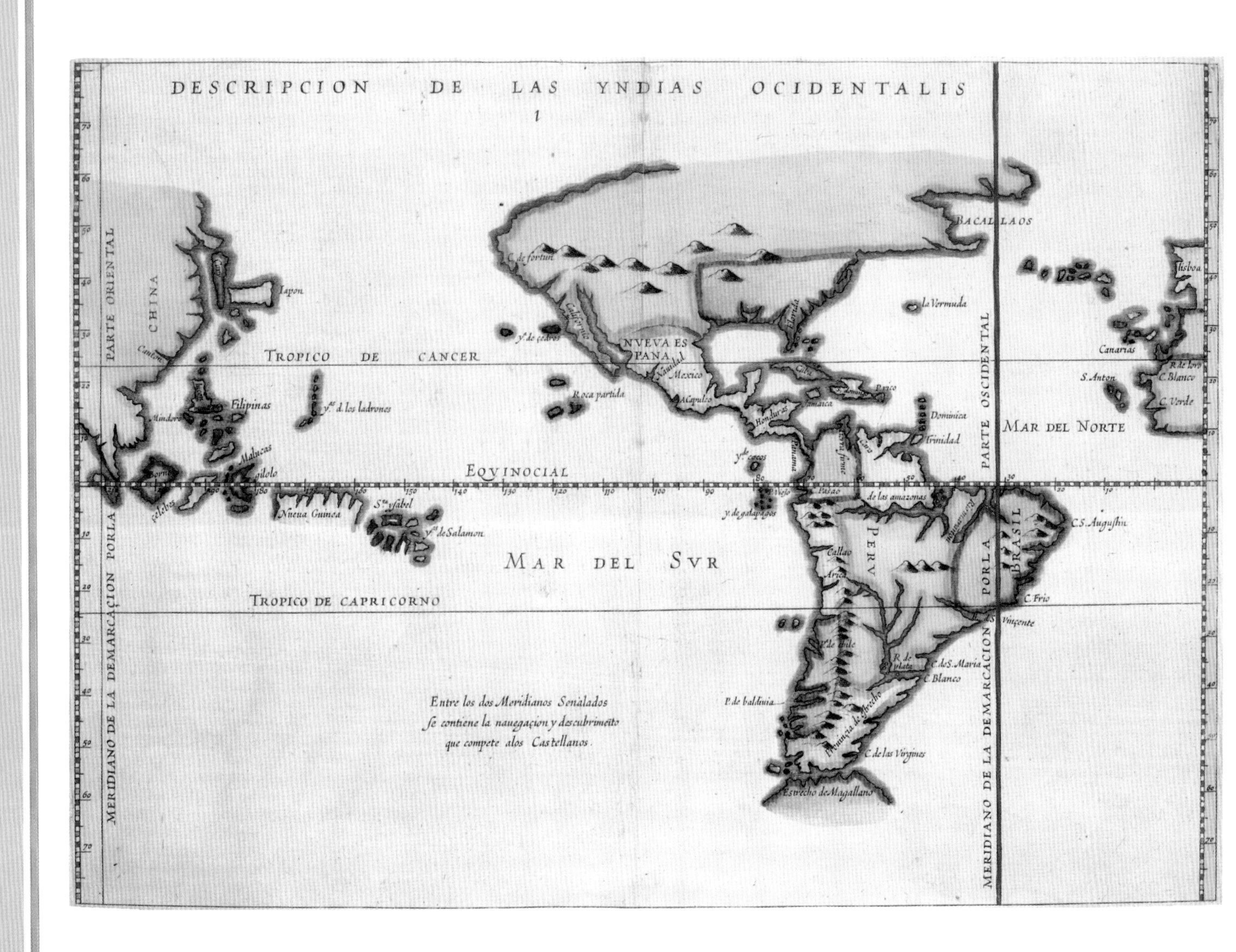

Yndias Ocidentalis—CA shown as attached land mass 1622 CA first thought to be attached.

By the time this depiction of *Las Yndias Ocidentalis,* or the West Indies, appeared in 1622, Europeans had a better grasp of New World geography. They understood, for instance, that the Amazon River rose amid the peaks of the Andes, that North America was bisected by a major river system—the Mississippi—and that California was not an island, but rather, part of the continent.

Explorer Father Kino's map (1743).

Between 1698 and 1701 Jesuit missionary Eusebius Francis Kino conducted several expeditions to the Gila River region of modern-day Arizona. From the heights near the junction of the Colorado and Gila rivers, Kino was able to look over into California, but he never actually entered it. Kino's explorations not only made him among the first to approach California by land but also demonstrated that the Gulf of California was closed at its north end. This proved conclusively that California could not be an island. The original of this map, published in England in 1743, is thought to have been drawn by Kino.

La Californie ou Nouvelle Caroline: teatro de los trabajos, Apostolicos de la Compa. e Jesus en la America Septe. par N. de Fer, Geographe de sa Majesté Catolique (c. 1720).

Despite the reports of Juan Rodríguez Cabrillo and later explorers such as Father Kino, the notion that California was an island persisted for centuries. This colorful map produced in Paris about 1720 shows that, even then, some mapmakers still did not understand that California was firmly attached to the North American continent. Here French mapmaker Nicholas de Fer depicts California as an enormous swordlike island separated from the continental landmass by a long, narrow strait. Details along the western shore of the island vaguely resemble those of the Baja California peninsula.

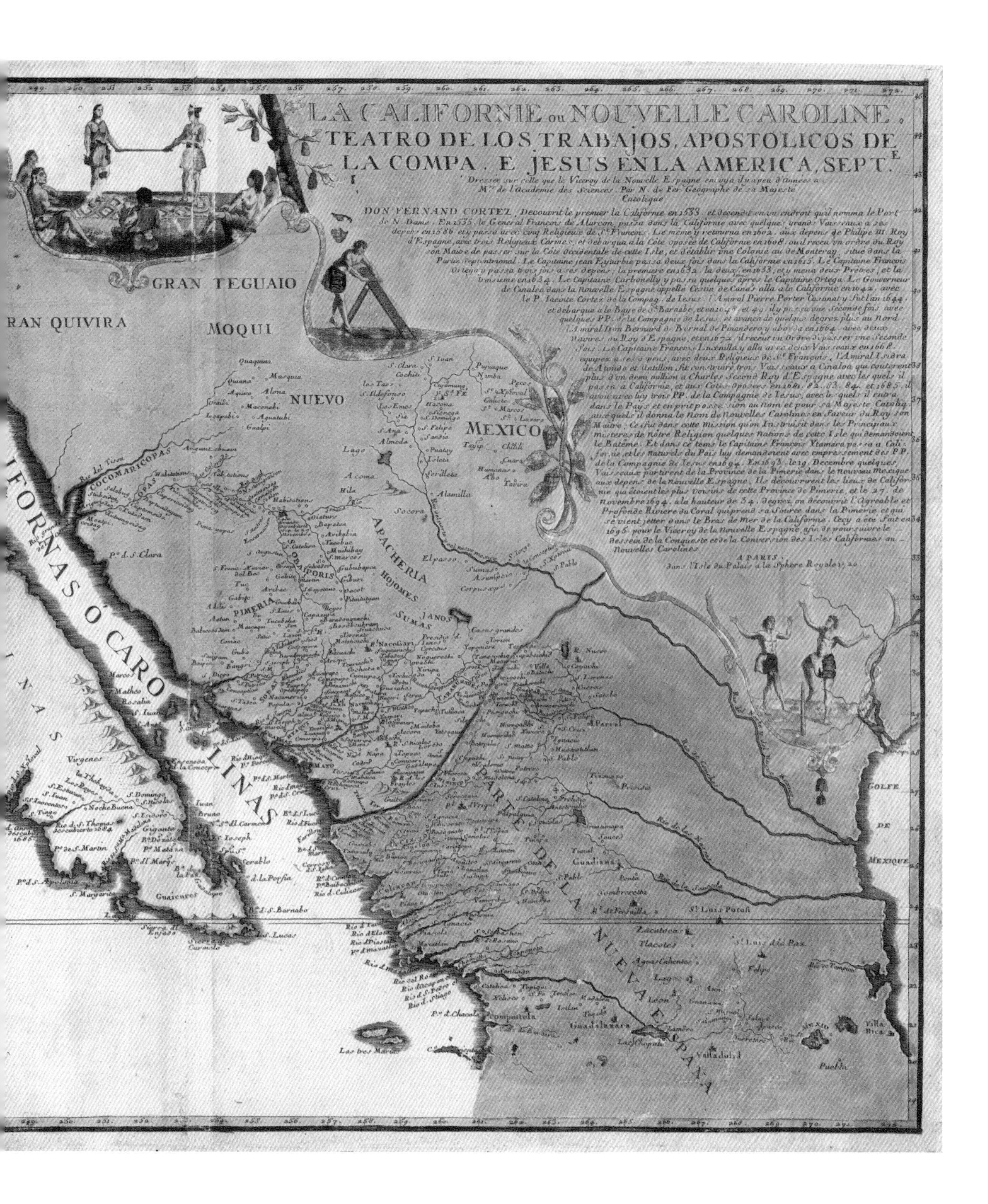
LA CALIFORNIE ou NOUVELLE CAROLINE.
TEATRO DE LOS TRABAJOS, APOSTOLICOS DE LA COMPA. E JESUS EN LA AMERICA, SEPT.E
GRAN TEGUAIO
RAN QUIVIRA
MOQUI
NUEVO MEXICO
APACHERIA
GOLFE DE MEXIQUE

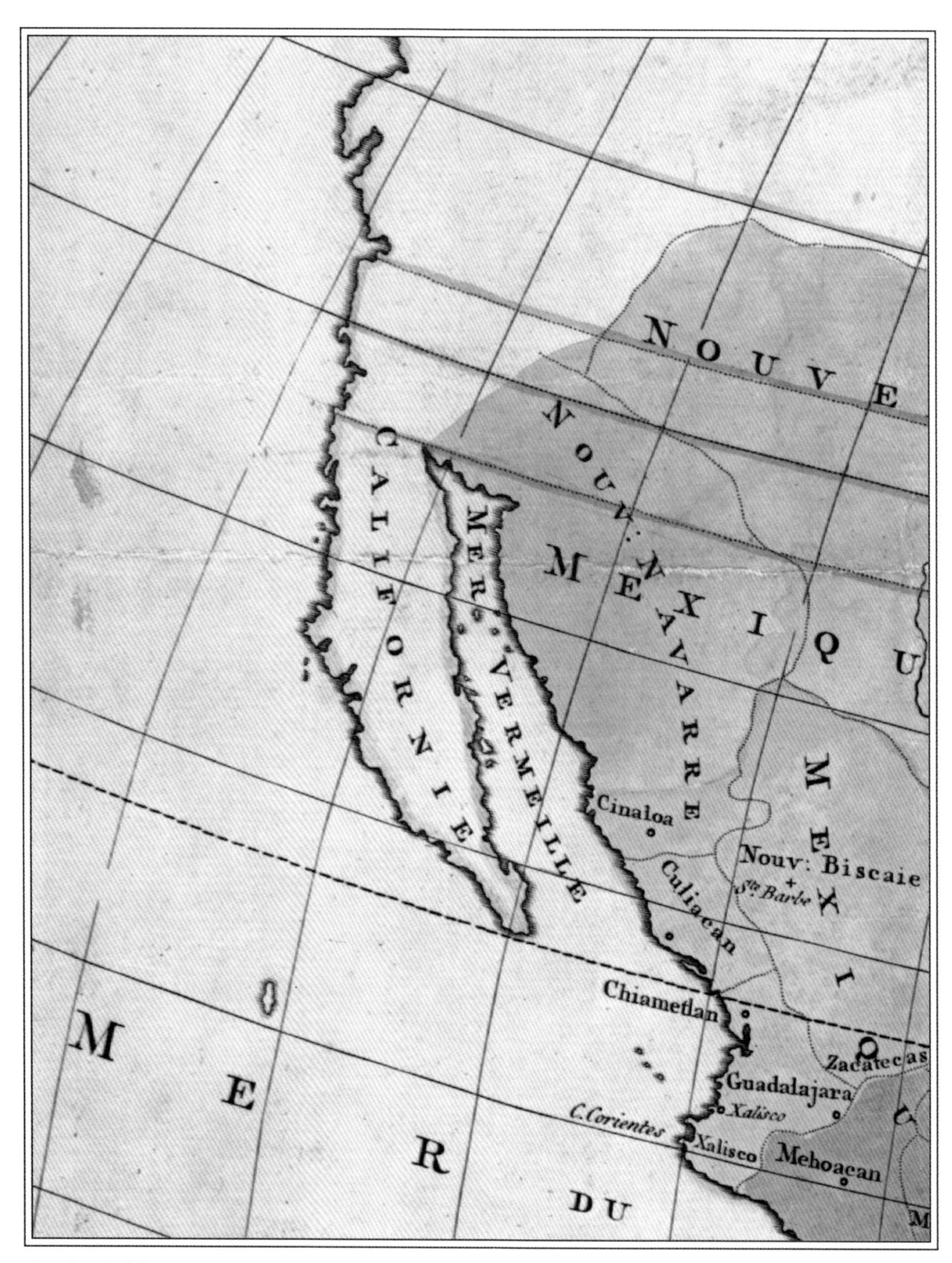

This detail of the map on pages 20–21 shows California as part of the North American continent.

Missions and Ranchos

For generations, California remained a wild and empty place, at least from the point of view of Europeans, and a highly tenuous part of the Spanish Empire. Spanish officials had their hands full attempting to settle and exploit the vast economic potential of Mexico and South America. They had little time and few resources to expend on a remote, little-known province located on the fringes of their domain. By the middle of the eighteenth century, however, Spanish authorities began to fear incursions into California by England, Russia, and other rival powers. In order to hold on to this potentially rich region, almost as large as Spain itself, they decided that California must be settled as quickly as possible. Their plan for accomplishing this was quite extraordinary, even if overly ambitious.

According to the plan, presidios or forts garrisoned by small contingents of Spanish troops would be established at strategic locations along the coast. Meanwhile, to provide California with a ready-made population of loyal Spanish subjects, local Native American peoples were to be converted to Christianity and turned into productive farmers. This gargantuan task was placed in the hands of a few dozen Franciscan friars who, under the leadership of Father Junipero Serra, were instructed to found a string of missions. Serra established the first of these missions in 1769 at San Diego and the second a year later at Carmel, near the new territorial capital of Monterey.

Serra envisioned a long line of similar wood, stone, and adobe missions strung out like rosary beads along a well-traveled connecting roadway. He believed these small pockets of civilization to be the seeds from which a settled Spanish province would grow. Such a transformation may have seemed impossible, perhaps even to Serra himself, but at the time he died in 1784 at Carmel, it appeared that his plan might be working. He had founded more than a dozen California missions himself, and many of them had become the core of self-sustaining, if not prosperous, farming communities. Although never much more than a dusty dirt track, Serra's roadway was established as well. Known as the Camino Real or Royal Highway, it would eventually stretch some 600 miles northward from San Diego.

The friars established no fewer than twenty-one separate missions along or near the Camino. The last of them was built in 1823 near Sonoma, a two-month journey on foot from San Diego. However, the missions stimulated little immigration from Spain or from other Spanish provinces. What is more, the Indians proved far less interested than Serra had hoped in becoming good Christian subjects of the Spanish Crown. While some California Indians did become Christians and a few adopted farming or skilled trades, most preferred their own traditional ways. By the time California became part of a newly independent Mexico in 1821, only a few thousand Spanish-speaking people lived in this vast territory. They relied for sustenance on a few farms, a scattering of cattle ranches, and the vineyards and plantations that had grown up around the missions.

Seeing that Serra's missions had failed in their original purpose of populating California, the Mexican government shut them down during the 1830s, evicting the friars and ordering them to put on their traveling sandals and trudge back to Mexico City. Mexican officials then turned to land grants as a means of civilizing the province. More than six hundred well-connected Mexicans and a few immigrants from other countries were given huge blocks of land that often spread across thousands of acres of rolling California hills.

These large rancho estates functioned much like feudal fiefdoms, usually under the tutelage of a single don and his extended family. The ranchos were used primarily for raising longhorn cattle that were periodically rounded up by vaqueros, some of the ablest and toughest cowboys ever seen in North America. Cattle hides were sold to sea traders from Europe or New England in exchange for fine cloth, china, and other luxury goods. Cowhides sold for about two dollars each, and the wealth they generated made it possible for the dons and their families to live a comfortable if not gracious existence in handsome haciendas. However, the ranchos would remain at the heart of California's economic and social life for only a few decades. Big changes were coming.

A colossus had risen in the East. Having survived its first few decades of existence and two great wars against the British, the United States was growing stronger and more populous by the day. The quarrelsome young nation seemed hell-bent on westward expansion, and it appeared likely the United States would eventually set its sights on California. The Louisiana Purchase in 1803 had nearly doubled the size of the country and brought U.S. borders hard against those of Spanish possessions in what is now the American Southwest. Ironically, Louisiana had formerly belonged to Spain but had been taken away by the French under Napoleon shortly before it was bought by President Thomas Jefferson and the United States for a mere $15 million, or about three cents an acre. Land-hungry settlers from the East soon poured into the vast Louisiana Territory, and inevitably, some of them kept going, ending up on the far side of the Sierra Nevada.

During the decades following its independence, Mexico maintained a relatively weak hold on California, so weak in fact that on more than one occasion Californians rebelled and tried to establish their own independent republic. One such attempt at succession was made in 1836 under the leadership of Juan Bautista Alvarado, an educated, native-born Californian from Monterey, the territorial capital. The Mexican government defused the situation by naming Alvarado governor and granting California a high degree of autonomy.

Mexico and Governor Alvarado attempted to discourage immigration by non-Spanish-speaking settlers. However, the trappers, prospectors, farmers, and ranchers who were just beginning to trickle over the mountains from the East proved largely unwilling to accept government by Mexico or to assimilate existing California culture. Their arrival in small but steadily increasing numbers during the late 1830s was rightly viewed as a threat to Mexican rule. It had long been believed that the waterless wastes of the Southwest and the near solid wall of the Sierras would block immigration in significant numbers from the East, but by the end of the decade entire wagon trains were getting through. Mexico had already had a very bad experience with just this sort of immigration in Texas, another remote and thinly populated territory. In Texas Anglo settlers from the United States had proven an unruly and ultimately ungovernable lot who in just a few years time managed to tear away a sizable chunk of Mexican territory and found their own republic.

Determined to prevent the same thing from happening in California, Mexican officials enlisted the assistance of Swiss adventurer John Sutter. Given a forty-eight-thousand-acre land grant along the western slope of the Sierras, Sutter built an impressive fort near the site of current-day Sacramento. It was hoped that Sutter and his fort would cork the leaky bottleneck of the Sierra passes. It did not. For one thing Sutter had his own agenda. He rapidly set about establishing what amounted to his own private kingdom, an entity he called New Switzerland. Since Sutter's domain needed commerce and a population, immigration, whether overland from the East or from elsewhere, could only promote his ambitions. Settlers continued to claw their way through the deserts and over the mountains. As it turned out, however, these bravest of souls would have little impact on events that were about to unfold in ways that neither the Mexicans nor Sutter could have anticipated.

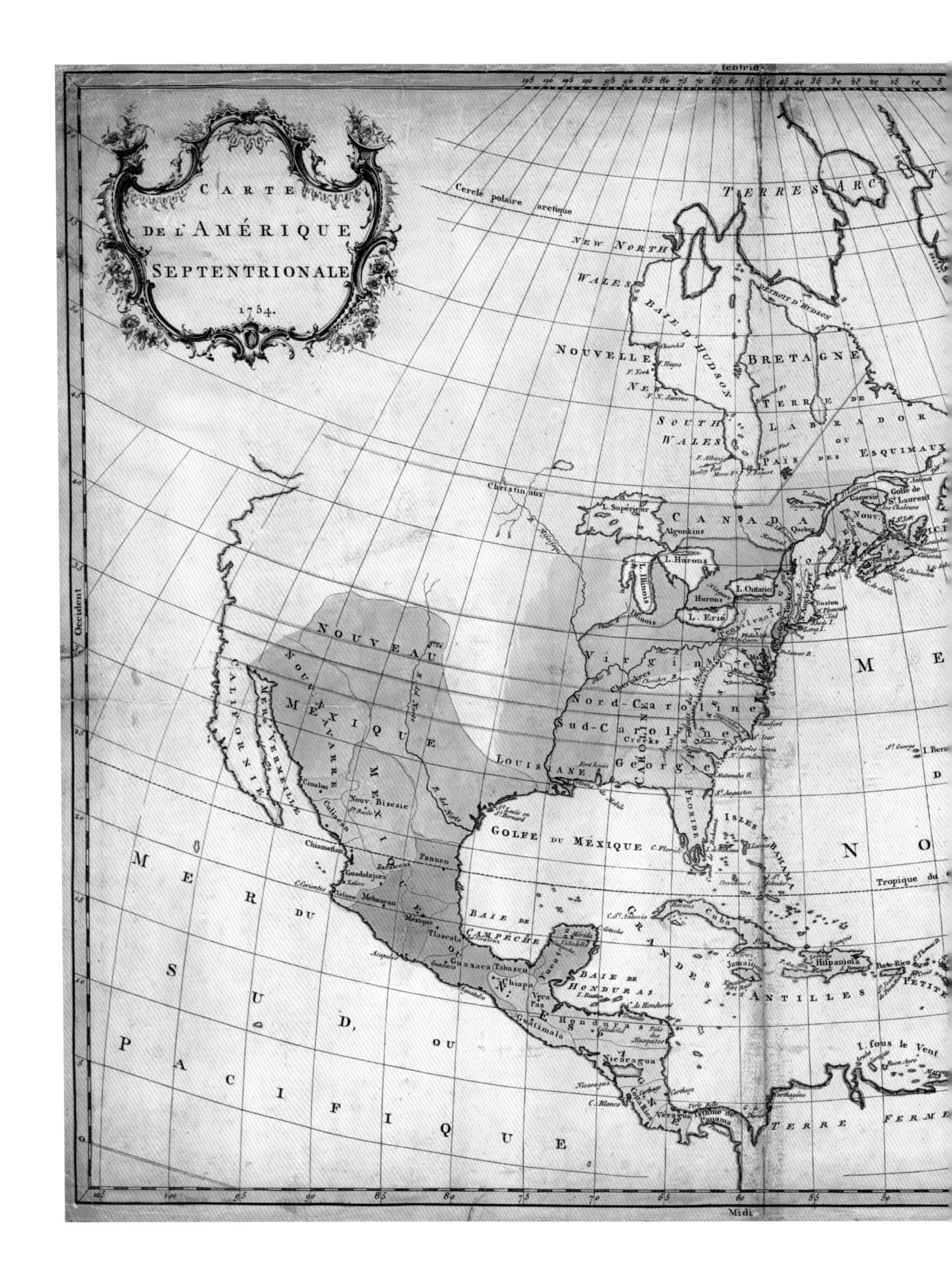
CARTE DE L'AMÉRIQUE SEPTENTRIONALE
1754.
Cercle polaire arctique
TERRES ARC
NEW NORTH WALES
BAIE D'HUDSON
DETROIT D'HUDSON
NOUVELLE BRETAGNE
NEW SOUTH WALES
TERRE DE LABRADOR OU PAIS DES ESQUIMAUX
Christinaux
L. Supérieur
CANADA
Algonkins
Golfe de St. Laurent
Quebec
L. Hurons
L. Illinois
L. Ontario
L. Erie
Hurons
Illinois
Boston
Long I.
NOUVEAU MEXIQUE
NOUV. MEXIQUE
CALIFORNIE
MER VERMEILLE
NAVARRE
Virginie
Cherakees
Nord-Caroline
Sud-Caroline
Creeks
CAROLINE
Georgie
LOUISIANE
FLORIDE
St. Augustin
Cinaloa
Nouv. Biscaie
MEXIQUE
GOLFE DU MEXIQUE
ISLES DE BAHAMA
Tropique du
Chiametlan
Panuco
Zacatecas
Guadalajara
Mechoacan
Mexique
Tlascala
BAIE DE CAMPÊCHE
Acapulco
Guaxaca
Tabasco
Chiapa
Yucatan
Vera Paz
BAIE DE HONDURAS
Honduras
Guatimala
Nicaragua
Cuba
Jamaique
Hispaniola
Porto-Rico
GRANDES ANTILLES
PETITE
I. sous le Vent
Panama
TERRE FERME
MER DU SUD, OU PACIFIQUE
M
N
O
Occident
Midi

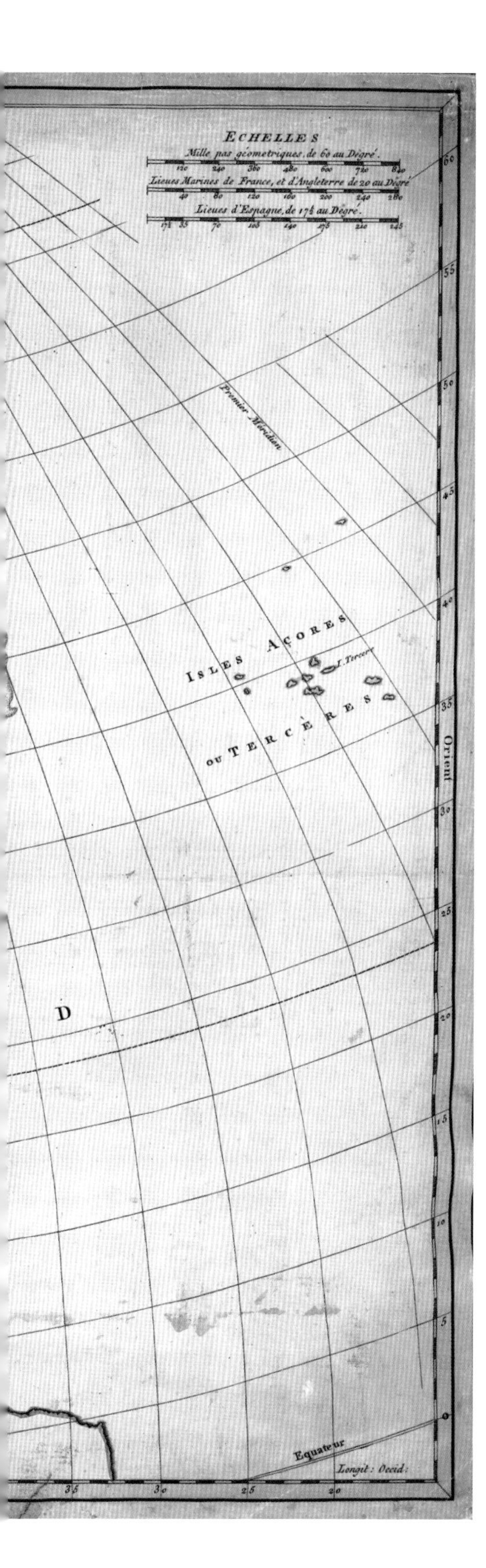

Carte de l'Amérique septentrionale (1754).

Part of an atlas published in 1755 by French cartographer Jean Palairet, this handcolored map shows boundaries and some cities, towns, and forts. Few if any such details are included for California, suggesting that, in the minds of Europeans at least, it remained on the fringes of the known world. At this time, however, the world was about to plunge into the Seven Years War, known in America as the French and Indian War. The Spanish, who fought unsuccessfully on the side of the French against Britain and Prussia, rightly feared that their enemies coveted Spain's possessions in the New World. Unsettled and thinly garrisoned, California seemed especially vulnerable.

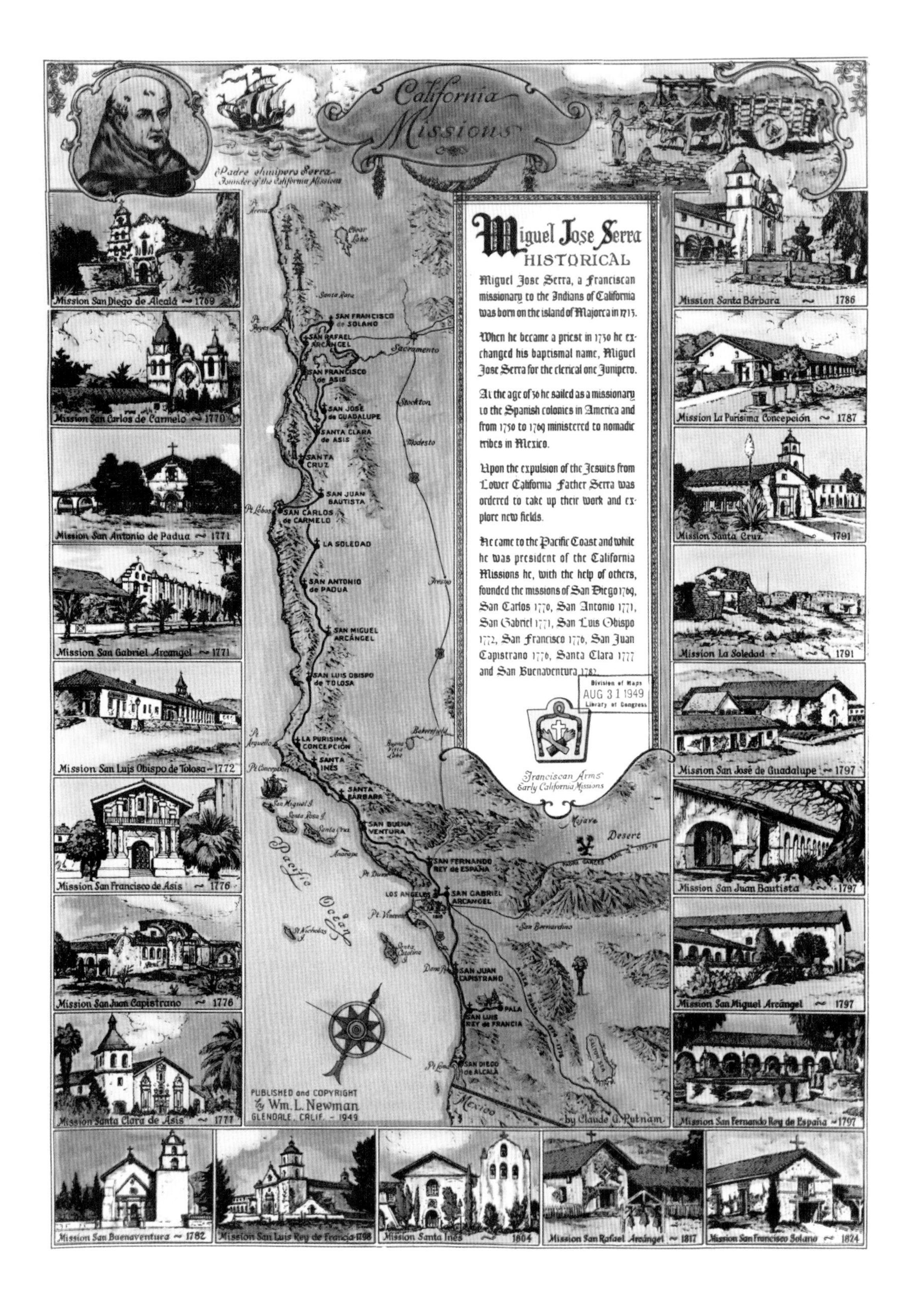
California Missions
Padre Junipero Serra
Founder of the California Missions
Miguel Jose Serra
HISTORICAL
Miguel Jose Serra, a Franciscan missionary to the Indians of California was born on the island of Majorca in 1713.
When he became a priest in 1730 he exchanged his baptismal name, Miguel Jose Serra for the clerical one Junipero.
At the age of 30 he sailed as a missionary to the Spanish colonies in America and from 1750 to 1769 ministered to nomadic tribes in Mexico.
Upon the expulsion of the Jesuits from Lower California Father Serra was ordered to take up their work and explore new fields.
He came to the Pacific Coast and while he was president of the California Missions he, with the help of others, founded the missions of San Diego 1769, San Carlos 1770, San Antonio 1771, San Gabriel 1771, San Luis Obispo 1772, San Francisco 1776, San Juan Capistrano 1776, Santa Clara 1777 and San Buenaventura 1782.
Division of Maps
AUG 31 1949
Library of Congress
Franciscan Arms
Early California Missions
Mission San Diego de Alcalá ~ 1769
Mission San Carlos de Carmelo ~ 1770
Mission San Antonio de Padua ~ 1771
Mission San Gabriel Arcangel ~ 1771
Mission San Luis Obispo de Tolosa ~ 1772
Mission San Francisco de Asis ~ 1776
Mission San Juan Capistrano ~ 1776
Mission Santa Clara de Asis ~ 1777
Mission Santa Bárbara ~ 1786
Mission La Purisima Concepción ~ 1787
Mission Santa Cruz 1791
Mission La Soledad 1791
Mission San José de Guadalupe ~ 1797
Mission San Juan Bautista 1797
Mission San Miguel Arcángel ~ 1797
Mission San Fernando Rey de España ~ 1797
Mission San Buenaventura ~ 1782
Mission San Luis Rey de Francia 1798
Mission Santa Inés 1804
Mission San Rafael Arcángel ~ 1817
Mission San Francisco Solano ~ 1824
SAN FRANCISCO de SOLANO
SAN RAFAEL ARCÁNGEL
Sacramento
SAN FRANCISCO de ASIS
Stockton
SAN JOSÉ de GUADALUPE
SANTA CLARA de ASIS
Modesto
SANTA CRUZ
SAN JUAN BAUTISTA
Pt. Lobos
SAN CARLOS de CARMELO
LA SOLEDAD
SAN ANTONIO de PADUA
Fresno
SAN MIGUEL ARCÁNGEL
SAN LUIS OBISPO de TOLOSA
Bakersfield
Pt. Arguello
LA PURISIMA CONCEPCIÓN
SANTA INÉS
Pt. Concepcion
SANTA BARBARA
SAN BUENA VENTURA
Pacific Ocean
Mojave
Desert
SAN FERNANDO REY de ESPAÑA
LOS ANGELES
SAN GABRIEL ARCANGEL
San Bernardino
SAN JUAN CAPISTRANO
PALA
SAN LUIS REY de FRANCIA
SAN DIEGO de ALCALÁ
Mexico
PUBLISHED and COPYRIGHT
by Wm. L. Newman
GLENDALE, CALIF. - 1949
by Claude G. Putnam

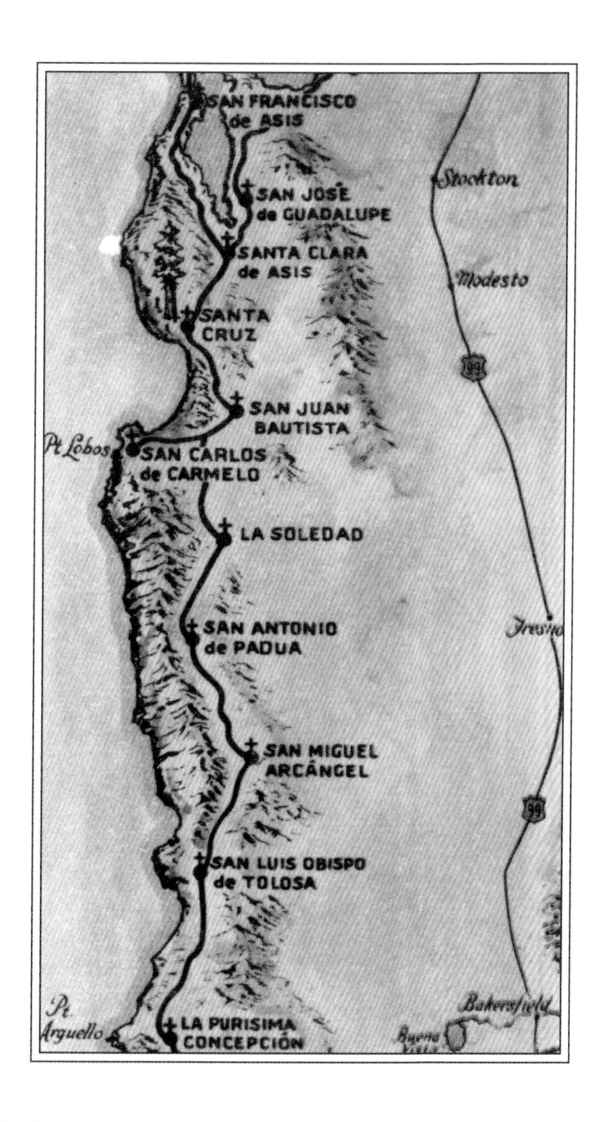

California missions (1949).

To protect California from covetous colonial powers in Europe, Spain sought to populate the province by converting Indians to Christianity and settling them on farms. This task was placed in the hands of Franciscan monks under Junipero Serra, who founded a chain of missions. The first was built in 1769 at San Diego and the last at Sonoma during the early 1820s, not long before a newly independent Mexican government shut the missions down for good. Although they failed in their original purpose, the missions are considered today an integral part of California's rich early history. This pictorial map, published in 1949, shows all twenty-one of the padre missions, along with the 600-mile Camino Real roadway that connected them. The detail above depicts a portion of the Camino Real linking Point Arguello to San Francisco as well as a modern highway traversing the Central Valley.

THE OLD
SPANISH AND MEXICAN
RANCHOS
OF LOS ANGELES COUNTY
MOJAVE DESERT
Los Alamos Y Agua Caliente
La Liebre
Temescal
San Francisco
Simi
Tujunga
1797
Ex-Mission de San Fernando
La Cañada
El Escorpion
El Encino
El Camino Real
Las Virgenes
Providencia
San Pascual
San Rafael
Santa Anita
Azusa de Duarte
Azusa Dalton
Addition to San Jose
El Conejo
Los Feliz
San Vicente Y Santa Monica
PUBLIC LAND
San Francisquito
San Jose
San Jose de Buenos Ayres
Rodeo de Las Aguas
La Brea
Pueblo de Los Angeles
San Gabriel Arcangel
1771
Potrero Grande
Potrero de Felipe Lugo
Los Nogales
Topanga
Malibu
Sequit
Boca de Santa Monica
Rincon de Los Bueyes
Las Cienegas
La Merced
Potrero Chico
La Puente
La Ballona
Cienega o' Paso de la Tijera
San Antonio
Paso de Bartolo
Rincon de La Brea
La Habra
Aguaje de La Centinela
Santa Gertrudes
Sausal Redondo
Tajauta
Los Coyotes
San Pedro
Los Cerritos
Los Alamitos
Los Palos Verdes
Prepared and Copyrighted by
Title Insurance
and Trust Company
Los Angeles
Scale of English Miles
Spanish Leagues
Spanish Varas

The old Spanish and Mexican ranchos of Los Angeles County, Gerald A. Eddy.

Since the missions had failed to provide California with a stable population, Mexican authorities turned to the establishment of ranchos as a means of civilizing the territory and tapping its rich economic potential. During the 1820s and 1830s, prominent ranchers were given vast land grants, which most used to raise longhorn cattle. The hides were sold to sea traders at prices high enough to support a comfortable lifestyle for the rancho dons and their families. However, the rancho system would not long survive the collapse of the hide market and the tide of immigration during the mid-1800s. This map, distributed by an insurance company during the 1930s, shows the location of historic ranchos in the Los Angeles area.

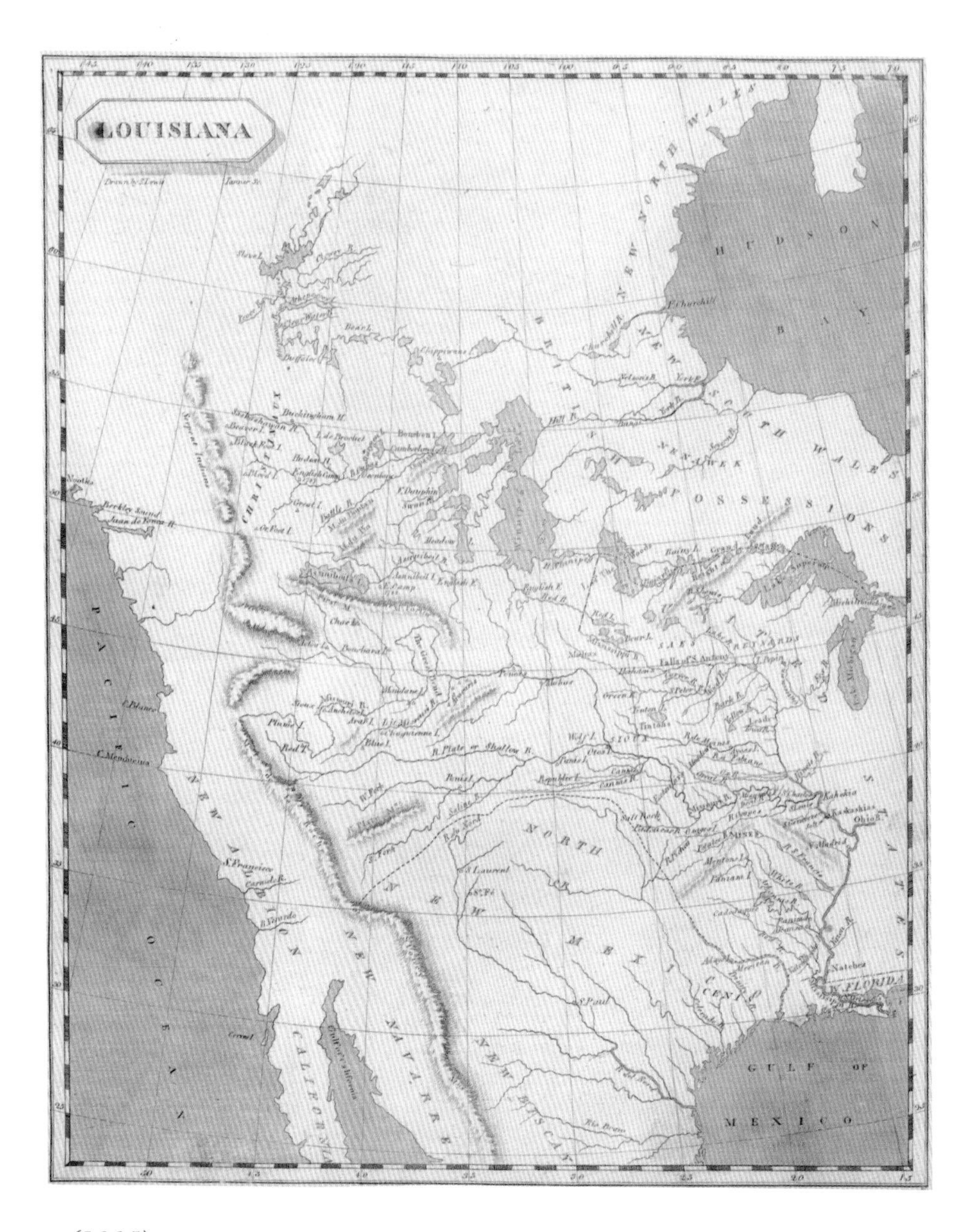

Louisiana (1805).

Ultimately, the most significant challenge to Spanish and later Mexican authority in California would come not from Europe but from the United States. This map was drawn by Samuel Lewis about 1805, after the United States had acquired the vast Louisiana Purchase from the French but before Lewis and Clark had completed their famous expedition to the Pacific. An extraordinary feature of the map is its positioning of the Rocky Mountains and the Continental Divide, far to the west of their actual location. This has the effect of diminishing the size and importance of Spanish possessions while seeming to bring the United States much closer to California.

A Map of the Internal Provinces of New Spain (1807).

An 1807 map attributed to explorer Zebulon Pike shows Spanish possessions in the Southwest. In 1806 Pike led an expedition to the Rocky Mountains, where he attempted but failed to climb the mighty peak that now bears his name. Pike pressed on into Spanish New Mexico and was arrested by authorities and held prisoner for several months. While in custody Pike had access to a treasure trove of Spanish maps, copies of which he carried home with him when repatriated during the summer of 1807. Interestingly, upper California is not prominently depicted on the map shown here.

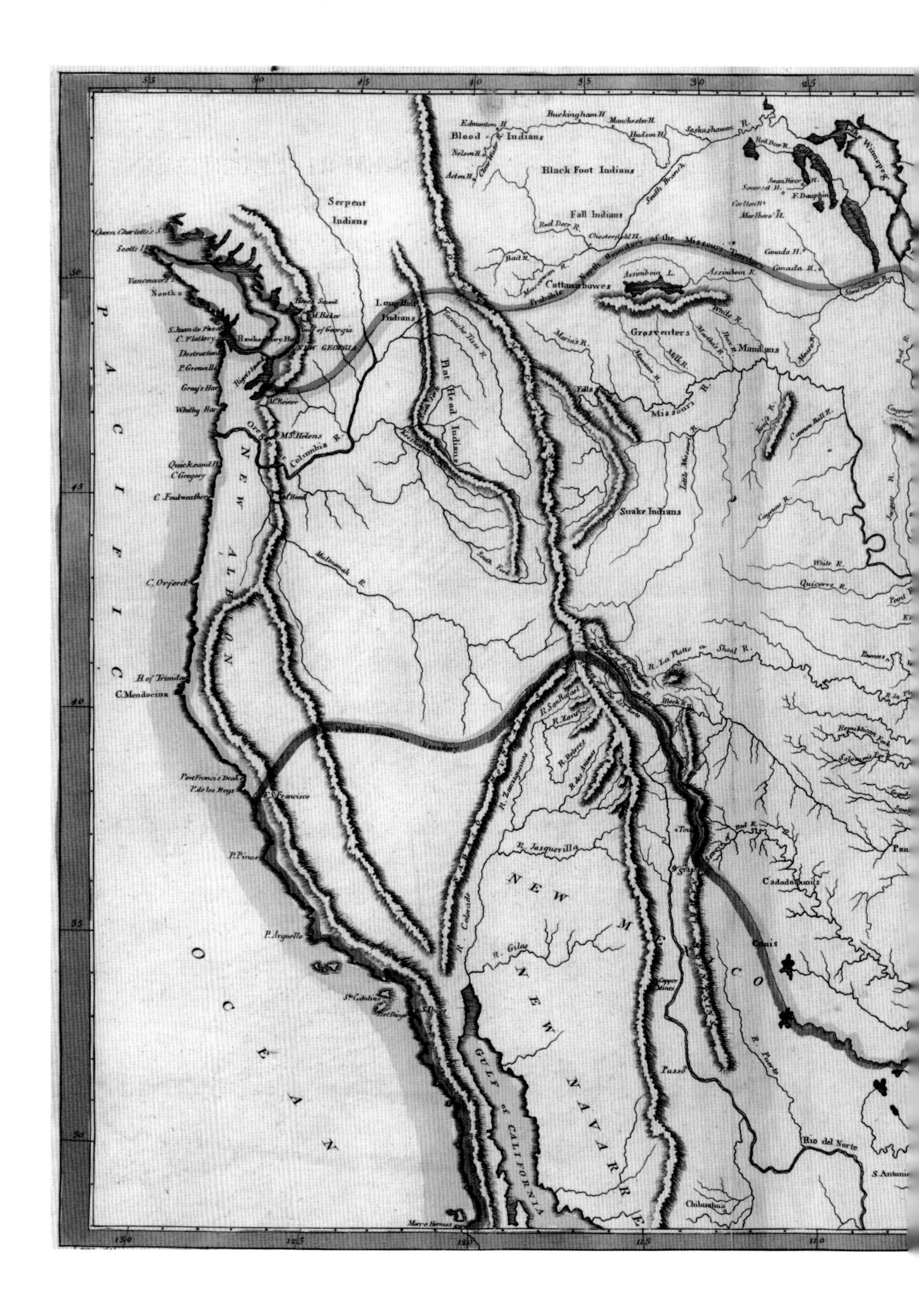

Edmonton H.
Blood Indians
Buckingham H.
Manchester H.
Hudson H.
Saskashawan R.
Nelson H.
Acton H.
Black Foot Indians
Serpent Indians
Fall Indians
South Branch
Red Deer R.
Chesterfield H.
Queen Charlotte's S.
Scotts I.
Vancouver's I.
Nootka
Bad R.
Probable North Boundary of the Missouri Territory
Canada H.
Assiniboin L.
Assiniboin R.
Cattanahowes
Grosventers
Mandans
Maria's R.
Milk R.
Missouri R.
Falls
Mt. Baker
Gulf of Georgia
New Georgia
S. Juan de Fuca
C. Flattery
Destruction I.
P. Grenville
Gray's Har.
Whitby Har.
Mt. Rainer
Mt. St. Helens
Columbia R.
Oregan
Long Hair Indians
Flat Head Indians
Quicksand B.
C. Gregory
C. Foulweather
Snake Indians
South Fork
Multnomah R.
Cannon Ball R.
Cayenne R.
White R.
Quicorre R.
C. Orford
R. La Platte or Shoal R.
B. of Trinidad
C. Mendocino
Pawnees
Republican Fork
Solomon's Fork
Probable Southern Boundary
R. San Rafael
R. Zaguananas
R. Dolores
Port Francis Drake
P. de los Reys
S. Francisco
R. Jasquerilla
P. Pinos
Cadodaquious
NEW MEXICO
P. Arguello
R. Colorado
R. Gilas
Copper Mines
Cenis
Sta. Catalina
S. Diego
Passo
PACIFIC OCEAN
NEW ALBION
NEW NAVARRE
GULF of CALIFORNIA
Rio del Norte
Chihuahua
S. Antonio
Mar o Bermeo

Missouri territory, formerly Louisiana (1814).
After the state of Louisiana joined the Union in 1812, the remainder of the original Louisiana Purchase was designated the Missouri Territory after the river of the same name. The Mathew Carey map shown here was created in 1814 while the United States was at war with Britain. Carey correctly depicts the Rockies and Sierras as two separate mountain chains. However, his map diminishes the relative size of the continent west of the Rockies. It also places the border between U.S. and Spanish territory, marked here in orange, well to the south of its actual location. Indeed, this map would appear to make San Francisco part of what would later become Oregon.

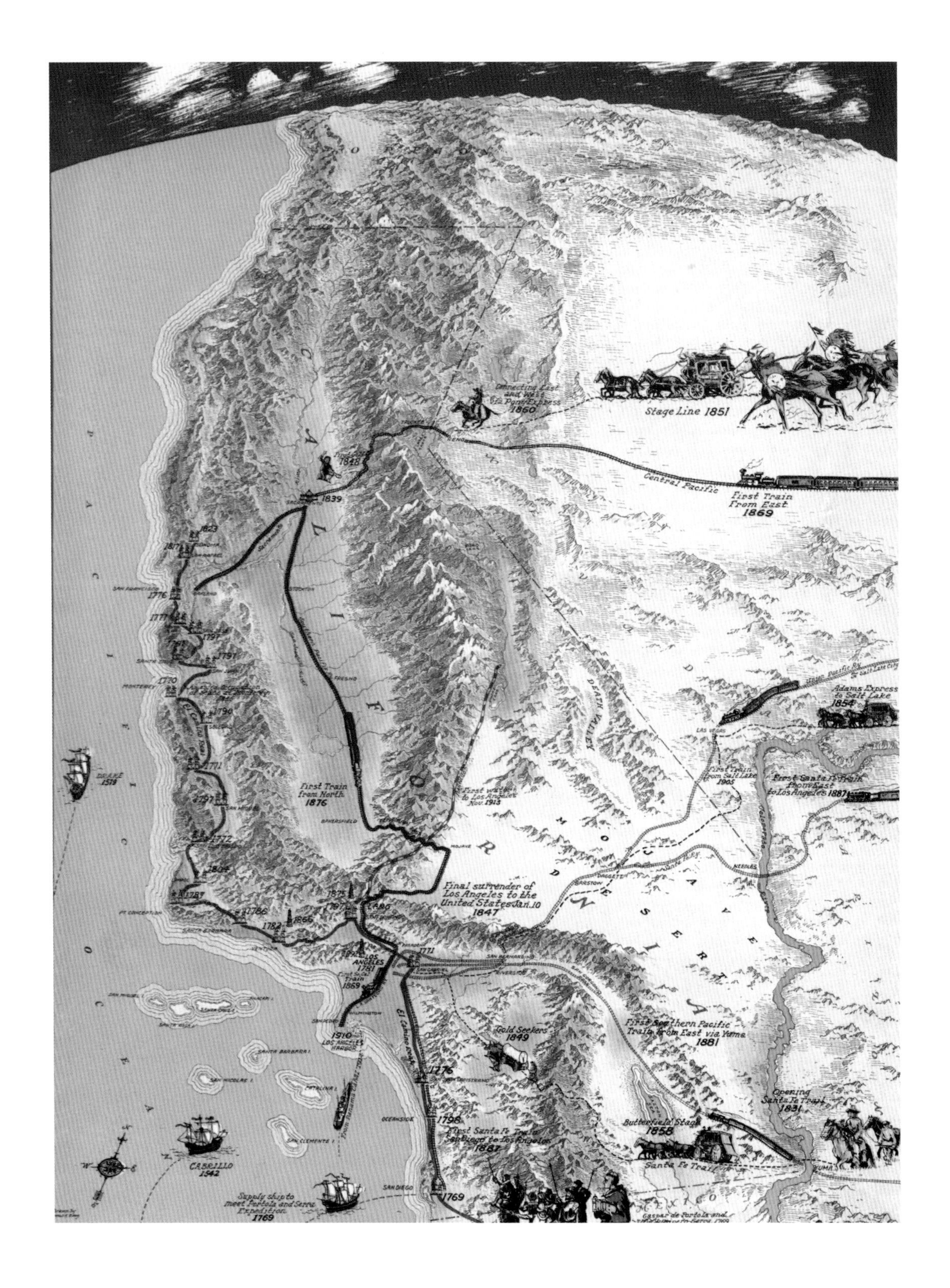
Stage Line 1851
1860
Central Pacific
First Train From East 1869
1839
Adams Express to Salt Lake 1854
DEATH VALLEY
First Train from North 1876
Final surrender of Los Angeles to the United States Jan.10 1847
First Southern Pacific Train from East via Yuma 1881
Gold Seekers 1849
Opening Santa Fe Trail 1831
Butterfield Stage 1858
Santa Fe Trail
CABRILLO 1542
Supply ship to meet Portola and Serra Expedition 1769
1769

Important Historical Events Which Have Made Los Angeles' Growth Possible, Gerald A. Eddy (1929).
An attractively drawn map of the Far West produced in 1929 illustrates various key events that contributed to the settlement of California and the Los Angeles basin. These include, among other important episodes, the exploratory expeditions, the establishment of the missions, migrations across the Sierra passes, the gold rush, and completion of the Transcontinental Railroad. History is a composite of forces and factors, and it may be that in California's case these led to a sort of delayed development. Because California settlement came late, it was especially dramatic and explosive.

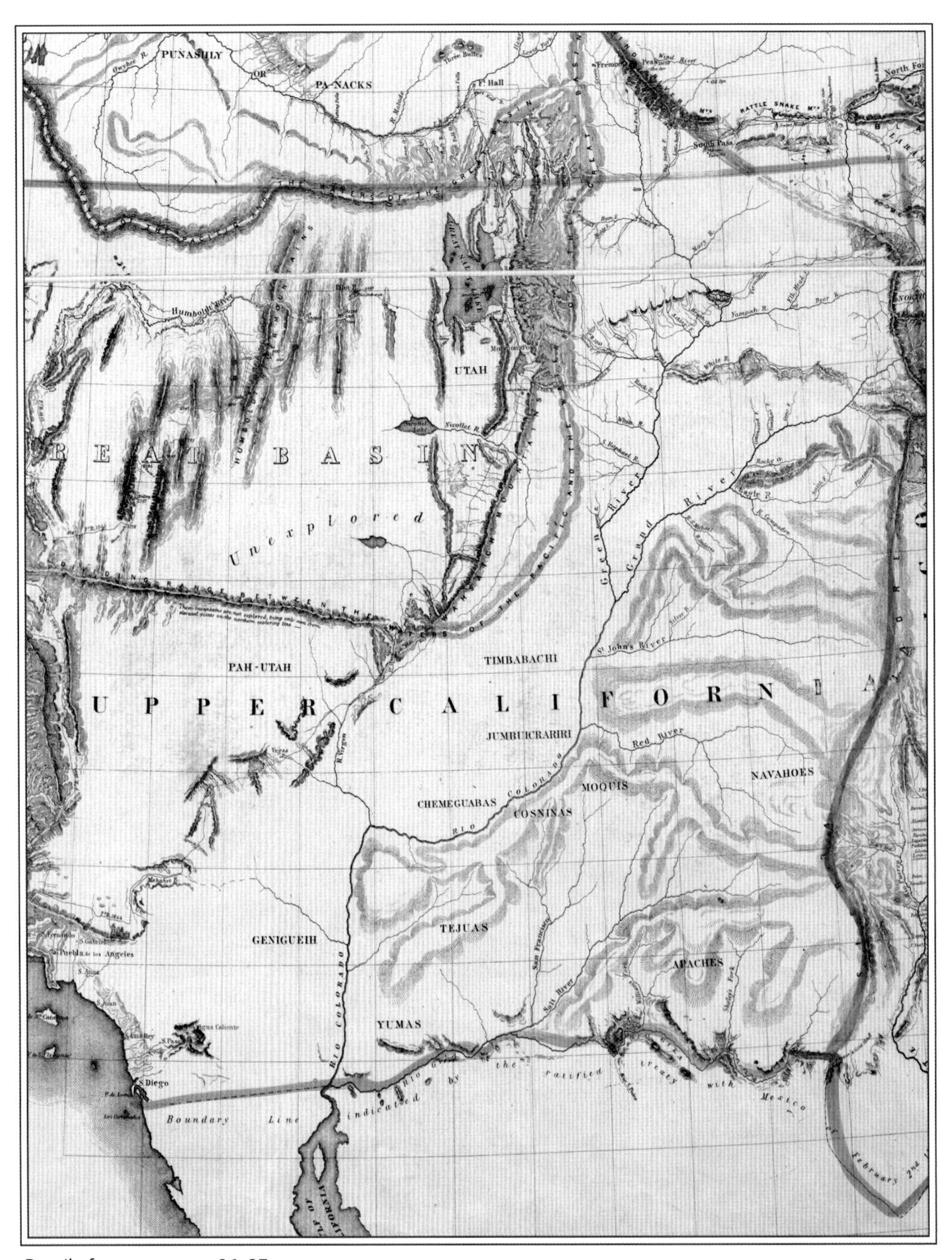

Detail of map on pages 36–37.

Gold Rush

After Texas was annexed by the United States in 1845, war between the United States and Mexico became more or less inevitable. Fighting broke out along the Rio Grande in the spring of the following year, and soon afterward war was declared by both sides. Though the news traveled quickly by early-California standards, it still took almost two months for word of war to reach California. After Mexico broke with Spain in 1821, almost a year passed before Californians learned that they were no longer subjects of the Spanish Crown. Though Mexican independence had seemed to be a momentous event, it changed little in the short run and came without the shedding of a single drop of blood in California. This time, however, Californians knew they were in for some hard fighting.

Cultural conflicts between English-speaking settlers and Spanish-speaking Californians had set the stage for conflict, and rumors of war between the United States and Mexico touched off an often-violent upheaval. It began on June 14, 1846, when thirty-four Sonoma settlers, fearing that they would be expelled from California, rebelled against Mexican authorities. Having stitched together a flag emblazoned with the crude image of a grizzly bear, they declared themselves guardians of the "Bear Flag Republic."

The revolt soon fell under the sway of Captain John Frémont, a U.S. Army engineer, who was supposed to be on a surveying and mapmaking expedition in Oregon. Hearing rumors of war, Frémont rushed southward into California where he threw his support behind the Bear Flag rebellion. Adding the sixty U.S. troops under his command to the tiny rebel army, Frémont was able to assemble a small but potent fighting force. Meanwhile, U.S. warships under Commodores John Sloat and Robert Stockton forced the bloodless capitulation of the presidios at Monterey and San Francisco. Thus, almost without firing a shot, Sloat, Stockton, and Frémont effectively secured Northern California for the United States.

However, many Californians still remained loyal to Mexico. They fought fiercely, especially in Southern California, where they bloodied U.S. troops at San Pedro, Rancho Dominguez, San Pasquel, La Mesa, and elsewhere, but ultimately their efforts were in vain. The fate of Mexican California would be sealed in the war's main theater, Mexico itself. Badly outnumbered, with few munitions industries and with little or no navy, Mexico could not hope to prevail against its much larger and wealthier foe. Predictably, Mexico's brave but underequipped armies fared poorly against the troops of U.S. general Winfield Scott in battle after battle. Mexico City fell shortly after the Battle of Chapultepec on September 13, 1847, and within months Mexican officials had signed the Treaty of Guadalupe Hidalgo. According to its terms, Mexico ceded approximately 500,000 square miles—about half its original land area—to the United States in exchange for a payment of $18.25 million. Part of the territory included in the settlement was California, and two and a half years later, California was admitted to the Union, with Frémont serving as one of the state's first U.S. senators.

During the war, John Frémont had taken possession of Sutter's Fort, renaming it Fort Sacramento, but otherwise the Mexican-American War barely touched Sutter's New Switzerland. Nor did it slow the influx into California of pioneer wagon trains from the East. One such wagon train, that of the so-called Donner Party, got stuck in the Sierra snows during the winter of 1846–1847. While General Scott was pounding Mexican defenses south of the Rio Grande, the eighty or so members of the Donner Party were eating through their supplies. When these were gone, they turned to cannibalism, cooking and eating their own frozen dead. By the time rescuers arrived in the spring, more than half the Donner Party had perished.

Originally intended to block unauthorized immigration from the East, Sutter's Fort ended up encouraging it instead. Settlers emerging from the Sierras often stopped at the fort to take on food and supplies. However, the relatively small and infrequent wagon trains that crossed into California during the early 1840s were nothing compared to what was coming. Ironically, during the closing days of the war, Sutter himself would unwittingly touch off one of the greatest mass migrations in history. During the fall of 1847, Sutter hired carpenter James Marshall and sent him to the American River to build a flour mill. On January 28, 1848, while checking the millrace for silt, Marshall's eye caught the glitter of metal. It turned out to be gold, and though Sutter swore Marshall to secrecy, word of the discovery soon got out, and the California Gold Rush was on.

No one is sure exactly how the madness began, but it is said that on or about May 10, 1848, a land speculator and newspaper publisher named Sam Brannan walked the streets of San Francisco waving a bottle of gold dust. "Gold!" he cried. "Gold has been discovered on the American River." Within a few days the city of San Francisco was a virtual ghost town. Everyone had packed up and rushed off to the Sierras to search for gold.

Passed along by newspapers, magazines, and word of mouth, news of the discovery spread like a California wildfire, not just in California itself but around the world. Even President James K. Polk got into the act, announcing to Congress in December 1848 that gold had been discovered in California. Soon passenger ships were arriving daily at the docks in San Francisco, loaded with amateur prospectors from the East, from Europe, and from China and Japan, all of them hoping to strike it rich in the mountains. They filled squalid camps in the Sierras, where they were forced to pay unheard-of prices for food, lodging, picks, and shovels and where most soon ran out of whatever money they had brought with them.

Busted prospectors, and there were many thousands of them, often worked as laborers for more successful miners, or they drifted back to San Francisco to look for work or practice a trade. More than a few found their true calling and made their fortunes outside the mining camps. For instance, Italian confectioner Domenico Ghirardelli found no gold in the Sierras, but he was eventually able to earn plenty of it by selling his delectable chocolate. A young Samuel Clemens, later known to the world as Mark Twain, joined the rush to California and, not finding any gold, turned to journalism. "I could not find honest work," Twain was fond of saying.

The majority of gold rush prospectors and immigrants arrived on ships. Some sailed across the Pacific, but most made the arduous 14,000-mile journey from the eastern United States around Tierra del Fuego at the extreme tip of South America and then northward to San Francisco. Many others booked passage to Panama, crossed the isthmus on foot or via mule trains, and then boarded ships bound for California. The gold rush peaked late in 1849 but would continue for several years. By the time it had played out, along with much of the original placer gold, during the mid- to late 1850s, upwards of three hundred thousand immigrants had flooded the state, and San Francisco had been turned, almost overnight, into a metropolis.

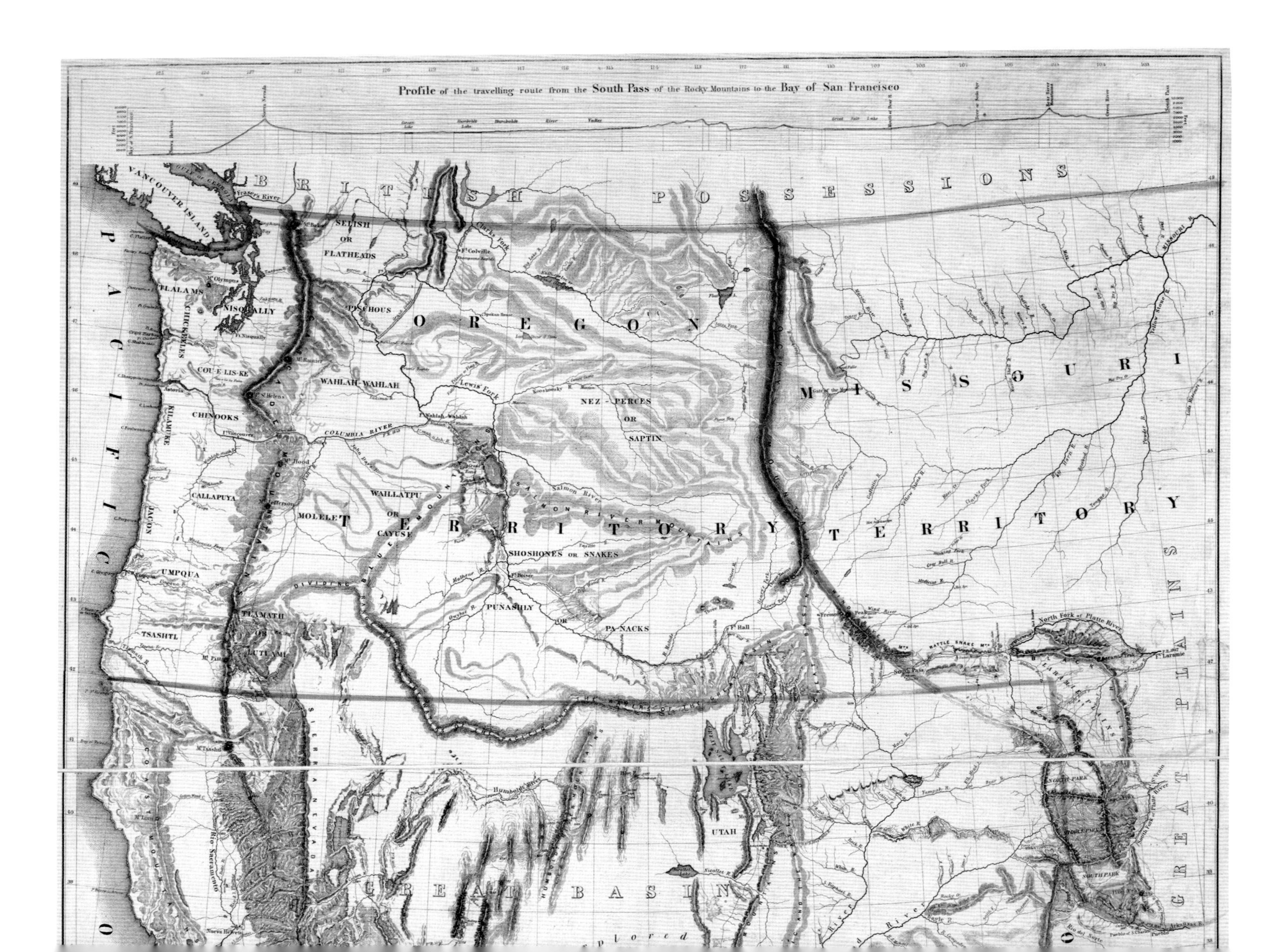
Profile of the travelling route from the South Pass of the Rocky Mountains to the Bay of San Francisco
BRITISH POSSESSIONS
VANCOUVER ISLAND
SELISH OR FLATHEADS
PISCHOUS
OREGON
MISSOURI
TERRITORY
ROCKY MOUNTAINS
NISQUALLY
CHICKEELES
COUE LIS-KE
WAHLAH-WAHLAH
NEZ-PERCES OR SAPTIN
CHINOOKS
KILAMUKE
COLUMBIA RIVER
CALLAPUYA
WALLATPU OR CAYUSE
MOLELET
JACON
SHOSHONES OR SNAKES
UMPQUA
PUNASHLY OR PANACKS
TLAMATH OR LUTUAMI
TSASHTL
SIERRA NEVADA
UTAH
GREAT BASIN
GREAT PLAINS
PACIFIC
Rio Sacramento
Salmon River
Lewis Fork
Clarks Fork
Humboldt River
North Fork of Platte River

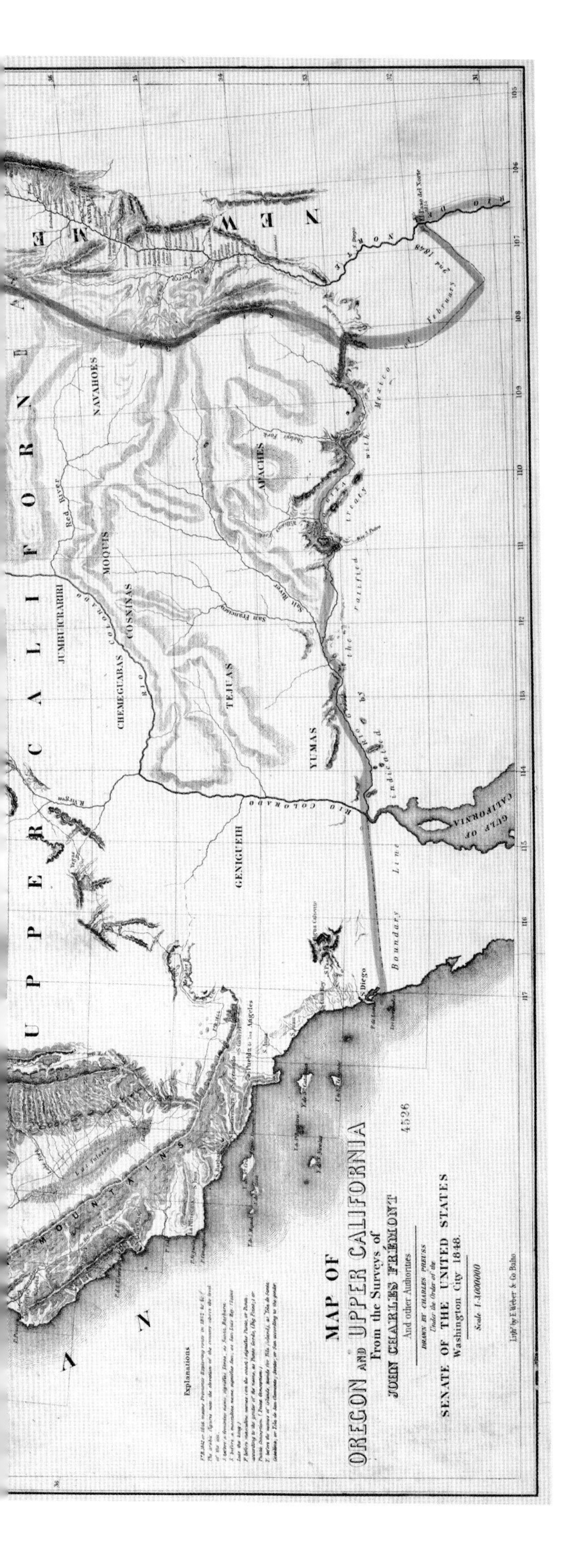

Map of Oregon and upper California from the surveys of John Charles Frémont and other authorities, drawn by Charles Preuss under the order of the Senate of the United States; lithy. by E. Weber & Co., Balto (1848).

Published by the U.S. Senate in 1848, the map shown here is based on surveys conducted by the explorer, soldier, and politician John Charles Frémont. Happening to be in the West when war broke out between the United States and Mexico in 1846, Frémont took charge of a rowdy force of U.S. immigrants already in rebellion against Mexican rule. While Frémont and other military leaders secured U.S. control of the interior, the U.S. Navy captured Monterey and San Francisco. Having lost several key battles, along with Mexico City, the Mexican government surrendered late in 1847 and early the next year ceded California and much of the Southwest to the United States in the Treaty of Guadalupe Hidalgo.

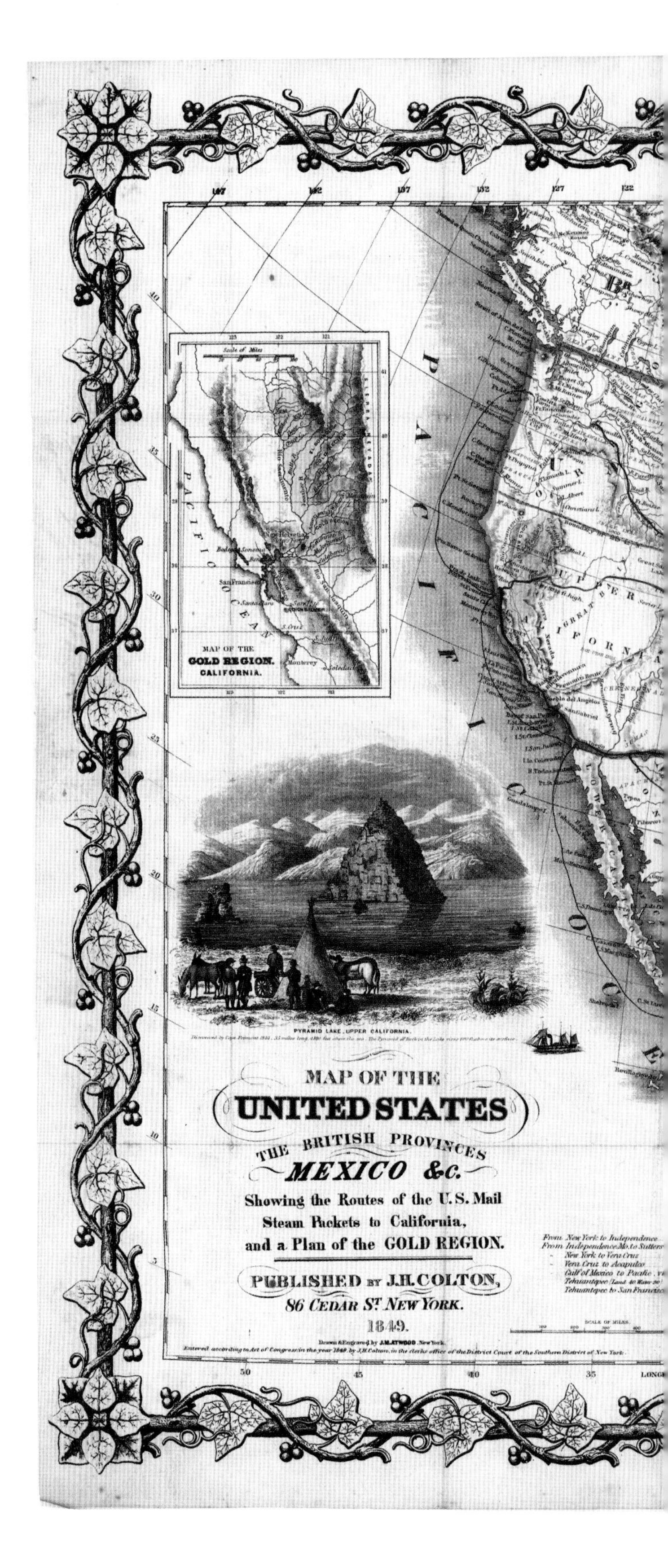

Map of the United States, the British provinces, Mexico &c. (1849).

By the time this map was published in New York in 1849, the United States had acquired about half of Mexico's former territory through the Treaty of Guadalupe Hidalgo. However, the border had still not reached its final position. The border as we know it today was drawn following the Gadsden Purchase, in which Mexico sold parts of present-day Arizona and New Mexico to the United States. No doubt map buyers in 1849 were primarily interested in the California gold country detailed in an inset on the upper left.

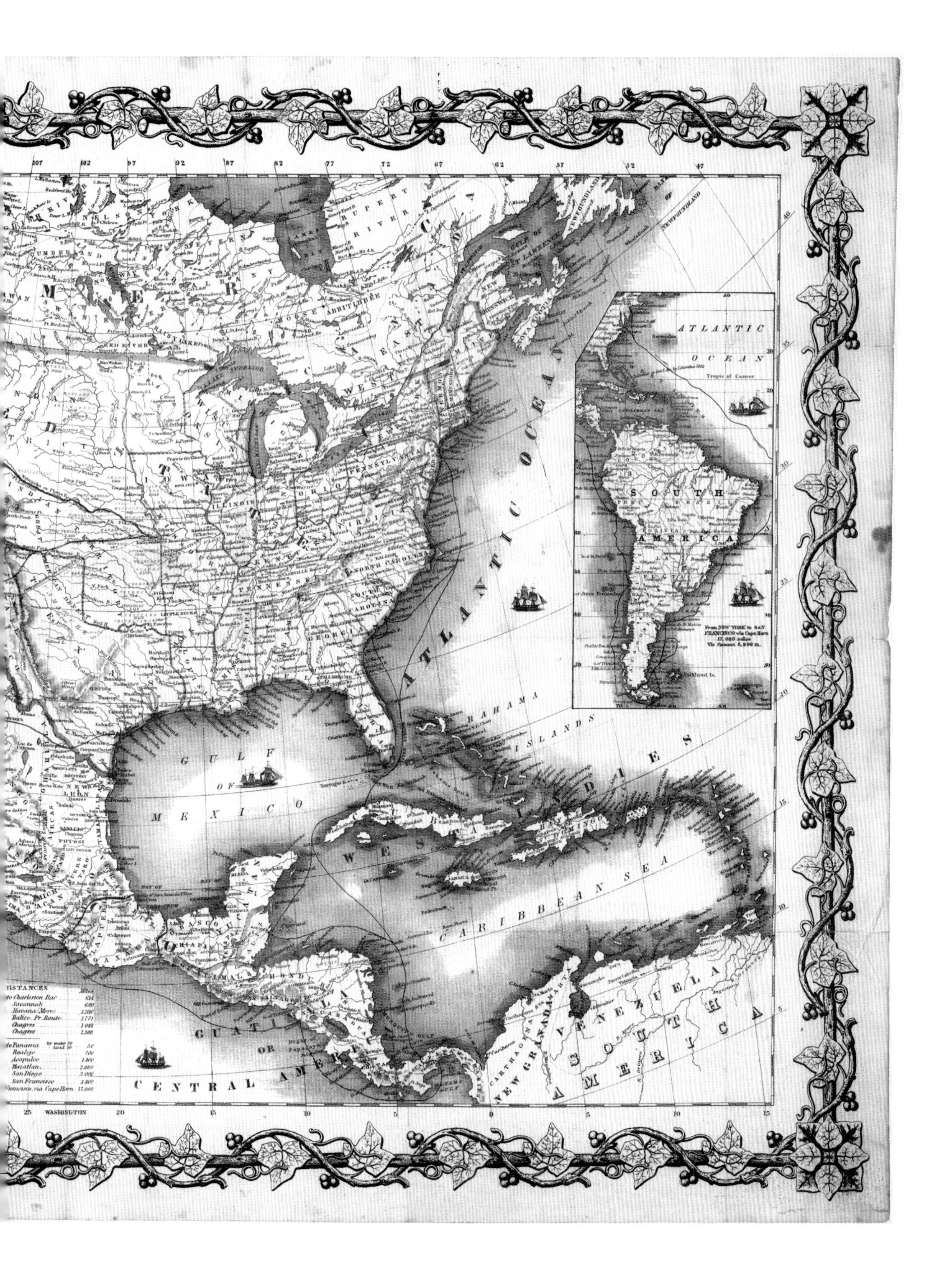
ATLANTIC OCEAN
GULF OF MEXICO
CARIBBEAN SEA
WEST INDIES
BAHAMA ISLANDS
SOUTH AMERICA
VENEZUELA
NEW GRANADA
CENTRAL AMERICA
GUATIMALA
NORTH CAROLINA
SOUTH CAROLINA
GEORGIA
TENNESSEE
ILLINOIS
OHIO
PENNSYLVANIA
VERMONT
NEW BRUNSWICK
NOVA SCOTIA
NEWFOUNDLAND
GULF OF ST LAWRENCE
ABBITIBBEE
RED RIVER
ATLANTIC OCEAN
Tropic of Cancer
SOUTH AMERICA
Falkland Is.
From NEW YORK to SAN FRANCISCO via Cape Horn 17,000 miles Via Panama 5,900 m.
DISTANCES
Miles
to Charleston Bar 614
Savannah 680
Havana (Moro) 1,200
Balize. Pr. Route 1771
Chagres 1,048
Chagres 2,308
to Panama by water 30 land 20 50
Realejo 700
Acapulco 1,800
Mazatlan 2,000
San Diego 3,000
San Francisco 3,600
Francisco, via Cape Horn 17,000
WASHINGTON

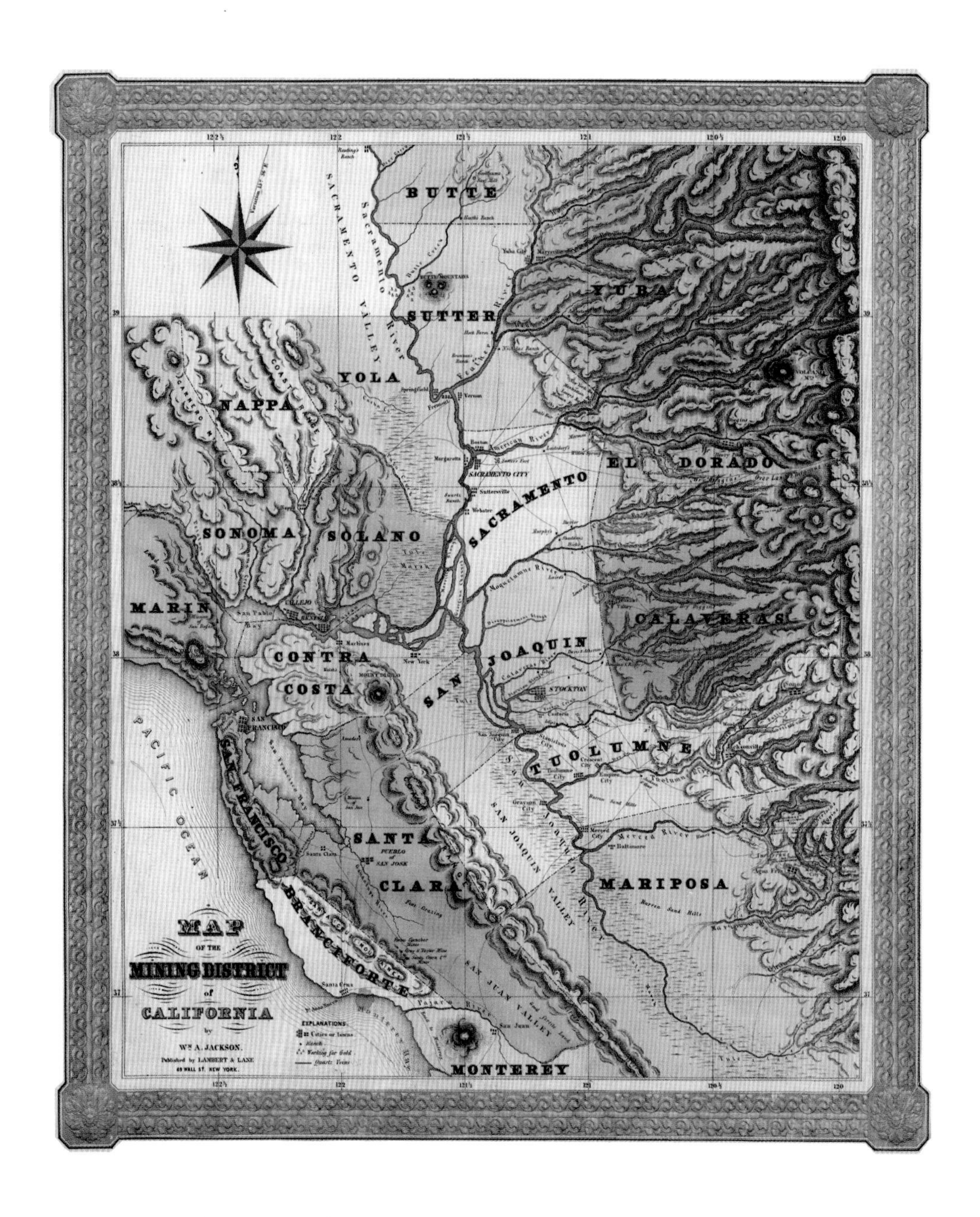

Map of the mining district of California.

It is easy to imagine would-be prospectors poring over maps like this one, trying to decide where they should try their luck. There really was a lot of luck involved in panning for gold and searching for a mother lode. Only one in a thousand prospectors ever came close to striking it rich, but those who did could found a family fortune on the proceeds—if they were careful. Many who had success in the goldfields soon gambled their riches away or lost their newfound wealth to swindlers.

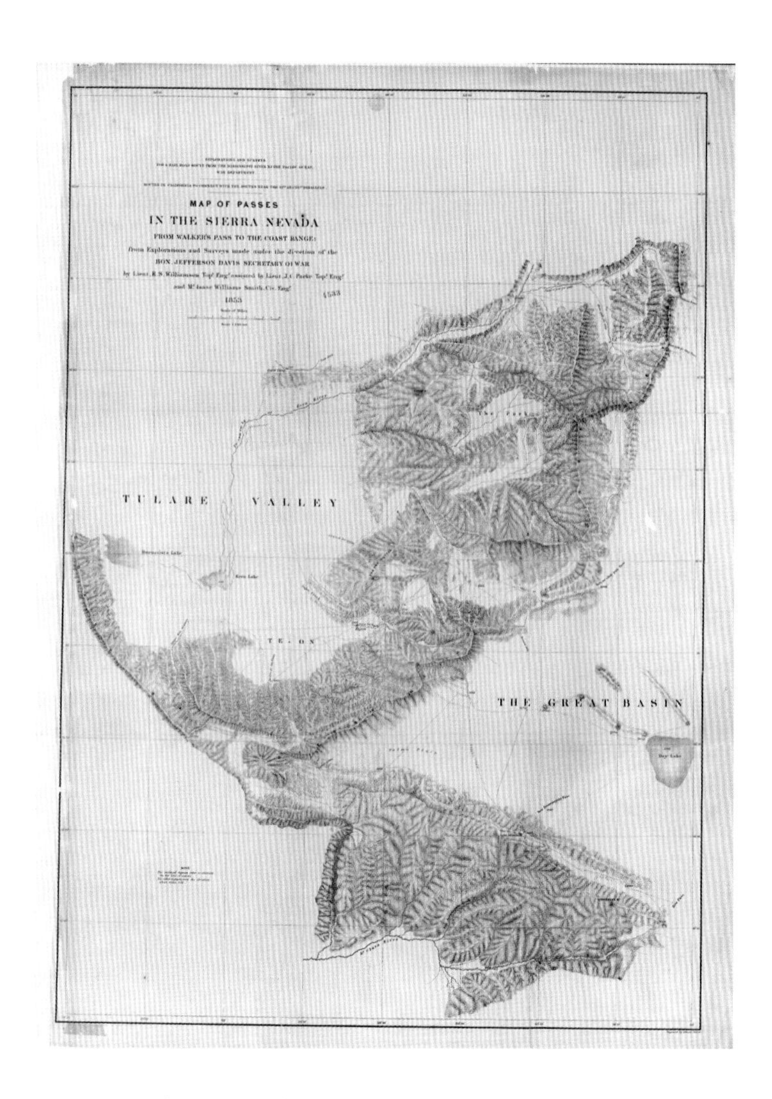

Map of passes in the Sierra Nevada from Walker's Pass to the Coast Range: from explorations and surveys, made under the direction of the Hon. Jefferson Davis, Secretary of War by Lieut. R. S. Williamson Topl. Engr. assisted by Lieut. J. G. Parke Topl. Engr. and Mr. Isaac Williams Smith, Civ. Engr. (1853).

While most early gold rush prospectors and immigrants traveled to California on ships, many arrived via overland routes from the East. This required crossing the rugged Sierras via one of several key passes, a difficult undertaking at any time and likely to be death dealing if attempted during the winter. Part of an extensive survey conducted during the 1850s by Secretary of War Jefferson Davis, this map shows mile-high Walker's Pass, about 150 miles north of Los Angeles. When the Civil War broke out in 1860, Davis became president of the Confederate States. Davis vainly hoped that an army from Texas might march on California and claim its goldfields for the South. As it turned out, the Confederates attempted only one invasion of the Far West, and it never got beyond New Mexico.

View of the Panamint Range Mountains, mines, mills and town site;
Sherman Town, property of the Panamint Mining & Concentration Works (1875).

Over the years fewer strikes were made and less gold ore was brought to the surface from played-out mines. By the late 1850s the gold rush had run its course, but there would be other California mineral booms. One of these occurred in the Panamint Mountains just west of Death Valley, where silver was discovered during the early

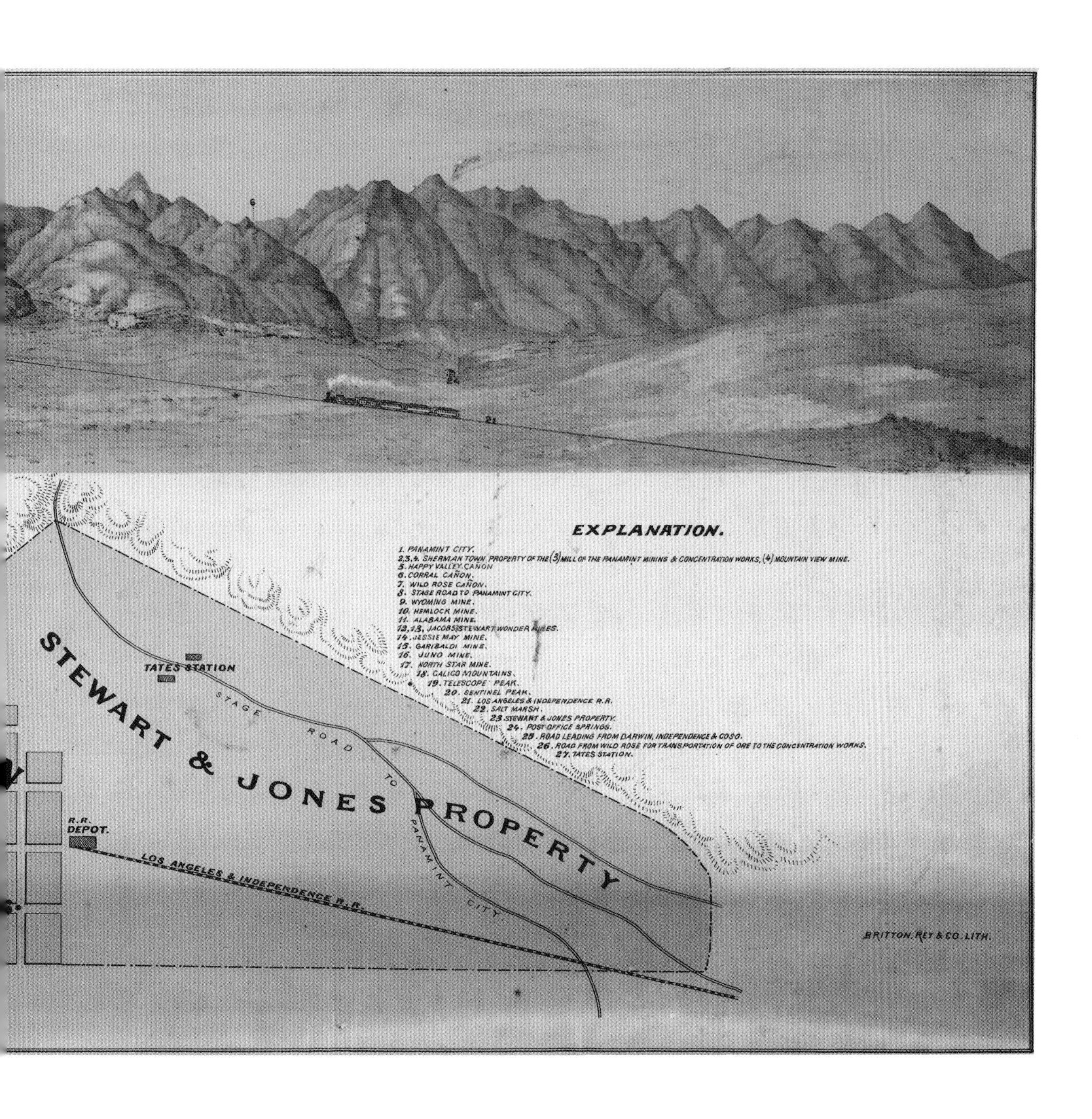

1870s. The Panamint strike produced so much silver that it was regularly cast into 450-pound slabs for shipment; the extraordinary weight of the ingots discouraged theft by bandits. Published in 1875, this map includes a view of the Panamint Range and of Sherman Town, also known as Panamint, a lawless place notorious for its saloons, red-light district, and frequent gunfights. Today Panamint is a ghost town.

OWENSMOUTH
SAN FERNANDO
VAN NUYS
LANKERSHIM
BURBANK
GLENDALE
LOS ANGELES
PASADENA
SIERRA MADRE
MONROVIA
AZUSA
GLENDORA
LA VERNE
SALT LAKE FIELD
MONTEBELLO FIELD
EL MONTE
PUENTE
POMONA
ONTARIO
SANTA MONICA
CULVER CITY
VENICE
HUNTINGTON PARK
WHITTIER
WHITTIER FIELD
PUENTE HILLS FIELD
WALNUT
CHINO
INGLEWOOD
WATTS
DOWNEY
SANTA FE SPRINGS FIELD
BREA CANYON FIELD
EL SEGUNDO
HAWTHORNE
ATHENS
LYNWOOD
STUDEBAKER
LA HABRA
OLINDA FIELD
BREA
MANHATTAN
LAWNDALE
NORWALK
COYOTE HILLS FIELD
FULLERTON FIELD
COMPTON
CLEARWATER
BELLFLOWER
RICHFIELD FIELD
HERMOSA
GARDENA
BUENA PARK
FULLERTON
REDONDO
DOMINGUEZ
ARTESIA
PLACENTIA
RICHFIELD
MID-OIL
HOME TRACT
REDONDO-TORRANCE FIELD
TORRANCE
SIGNAL HILL
FIELD
ANAHEIM
CYPRESS
LOS ALAMITOS
STANTON
LONG BEACH
WILMINGTON
ORANGE
GARDEN GROVE
WESTMINSTER
SANTA ANA
SAN PEDRO
SEAL BEACH
HUNTINGTON BEACH FIELD
SUNSET BEACH
PACIFIC OCEAN
FORNCROOK

LANKERSHIM
SALT LAKE FIELD
LOS ANGELES
PASADENA
SIERRA MADRE
MONROVIA
AZUSA
COVINA
GLENDORA
CLAREMONT
POMONA
ONTARIO
MONTEBELLO FIELD
WHITTIER
FIELD
PUENTE HILLS
FIELD
SANTA MONICA
VENICE
CULVER CITY
SANTA FE SPRINGS FIELD
BREA FIELD
CHINO
INGLEWOOD
DOWNEY
LA HABRA
OLINDA FIELD
HUNTINGTON-
CLEARWATER
SUBDIVISION
BREA
EL SEGUNDO
HAWTHORNE FIELD
NORWALK
COYOTE HILLS FIELD
FULLERTON FIELD
RICHFIELD FIELD
COMPTON
CLEARWATER
BUENA PARK
FULLERTON
PLACENTIA
RICHFIELD
HERMOSA
REDONDO
ARTESIA
ANAHEIM
TORRANCE FIELD
TORRANCE
CYPRESS
SIGNAL HILL FIELD
LOS ALAMITOS
STANTON
ORANGE
LONG BEACH
WILMINGTON
GARDEN GROVE
SANTA ANA
WESTMINSTER
HUNTINGTON BEACH FIELD
SAN PEDRO
SEAL BEACH
HUNTINGTON-
CLEARWATER
SUBDIVISION
SUNSET BEACH
PACIFIC OCEAN
HUNTINGTON BEACH
FORNCROOK

Views of oil fields around Los Angeles, C. S. & E. M. Forncrook (1922). The strike-it-rich mentality characteristic of the gold rush gripped Californians once again during the late nineteenth century, when black gold, also known as oil, was discovered practically beneath the streets of Los Angeles. At one time there was barely a spot in Los Angeles County where a person could stand and not see an oil derrick. By the 1920s Southern California ranked among the richest oil-producing regions in the world. These three views dating to 1922 show oil fields in and around Los Angeles.

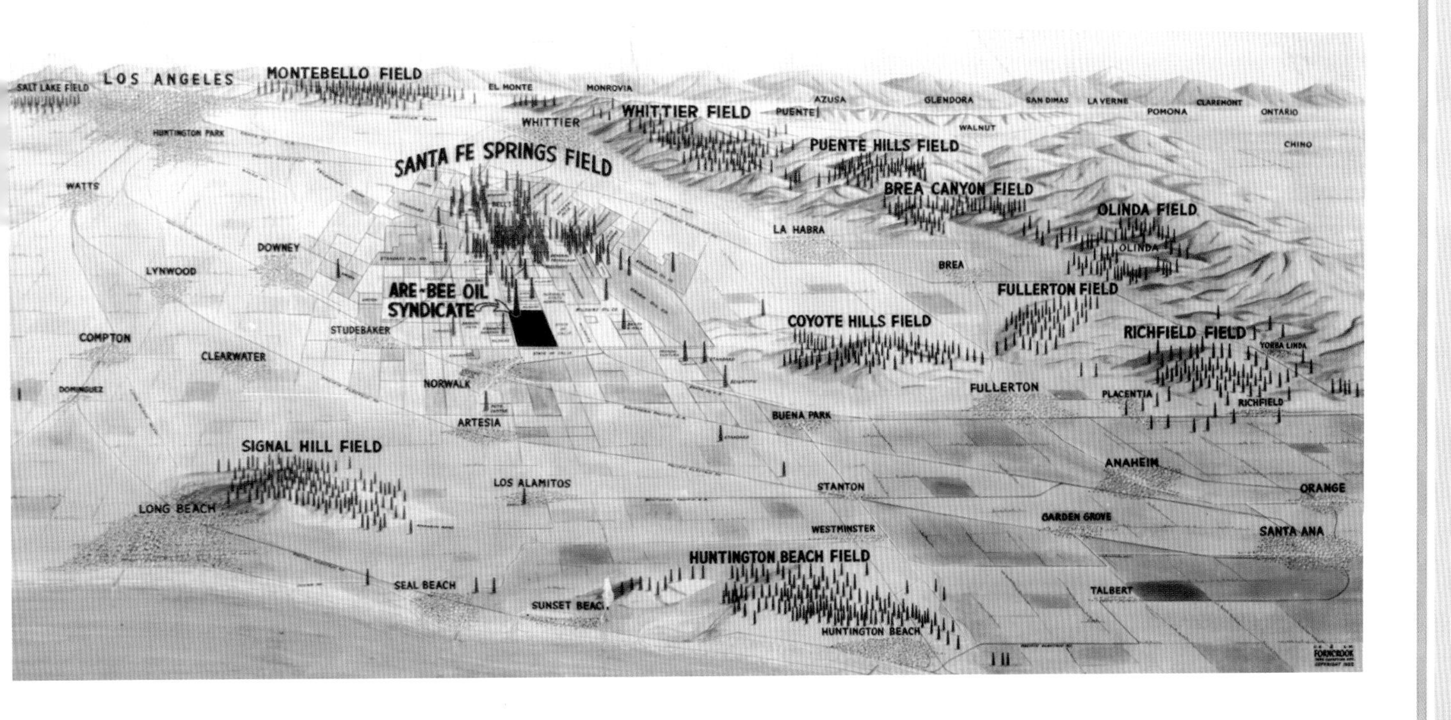

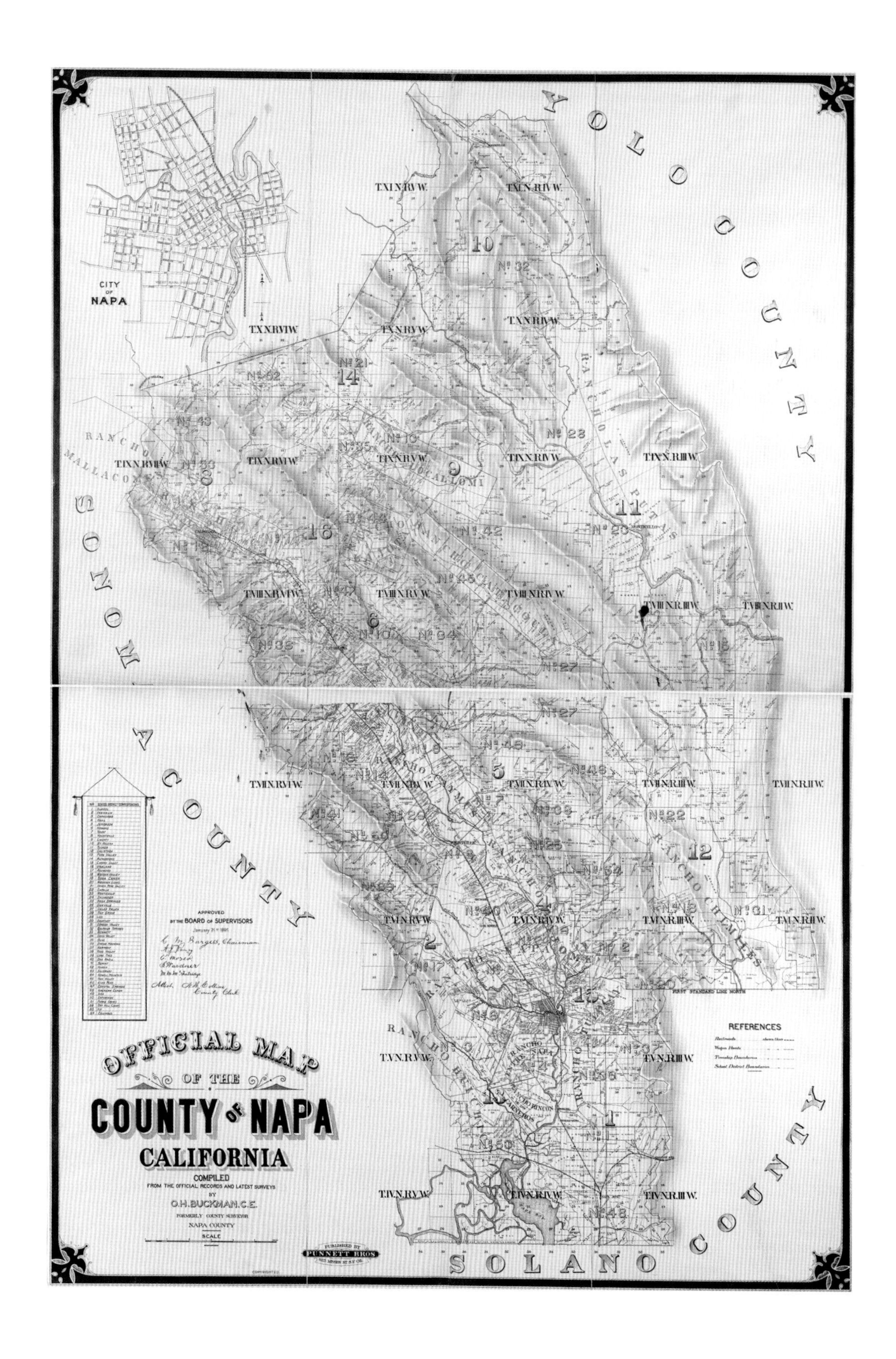

OFFICIAL MAP OF THE COUNTY OF NAPA CALIFORNIA
COMPILED FROM THE OFFICIAL RECORDS AND LATEST SURVEYS
BY
O.H. BUCKMAN, C.E.
FORMERLY COUNTY SURVEYOR
NAPA COUNTY
SCALE
PUBLISHED BY PUNNETT BROS
CITY OF NAPA
YOLO COUNTY
SONOMA COUNTY
SOLANO COUNTY
APPROVED BY THE BOARD OF SUPERVISORS
REFERENCES
FIRST STANDARD LINE NORTH

Official map of the County of Napa, California: compiled from the official records and latest surveys, by O. H. Buckman (1895).

Although silver was discovered in the Napa hills during the 1850s, the county played only a minor role in the gold rush. At the time this map was published in 1895 by the Punnett Brothers Company in San Francisco, Napa was a decidedly rural county relying almost entirely on agriculture. In fact, it remains highly agrarian to this day. However, during the mid to late twentieth century, Napa became the center of a new sort of boom focusing on the production of high-quality wines. Napa wines are ranked alongside the best in the world, and they attract prices that make living standards—and the quality of life—in this Northern California county among the highest in America.

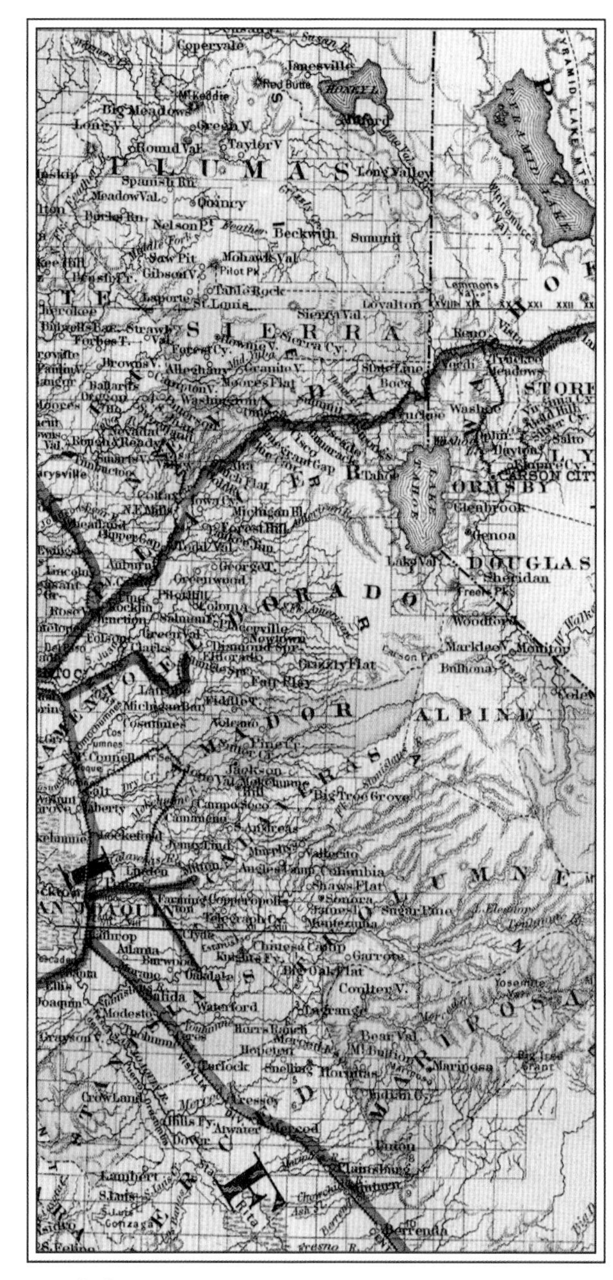

Detail of map on page 52.

Railroads and Transportation

Transportation was a major concern for California from the start. Lack of adequate transportation had retarded development as far back as the time of the padres and continued to do so well past the gold rush era. During the late eighteenth and early nineteenth centuries, the Camino Real had linked a few towns and villages and the twenty-one missions founded by the Franciscan padres, but it was never much more than a crude frontier trail. Important events could take place in Europe, Mexico, or the eastern United States, and people in California might not hear of them for a year or more. It could take ships eight or nine months to reach California from the North American East Coast and wagon trains almost as long to reach the western slope of the Sierras from Missouri. These journeys were not just long and arduous but dangerous as well. Many, like the less-fortunate members of the Donner Party, never made it to golden California.

The wealth generated by the gold rush made it possible to build a network of dirt roads. Wagons and stagecoaches bumped and rattled over these in ever-growing numbers, but a faster, more-reliable transportation system was sorely needed. A bold and relatively new nineteenth-century technology would provide the solution—railroads. The first steam railroads in the East had been built during the 1820s, but decades would pass before California got railroads. Naturally enough, the state's first railroad was intended to serve gold rush miners. Completed in February of 1856, the 22-mile Sacramento Valley Railroad linked the gold-mining country around Folsom with Sacramento, the state's new capital city. The little railroad's first locomotive was a twenty-five-ton, smoke-belching behemoth appropriately named *Elephant*.

Ironically, *Elephant* and the other locomotives, rails, and rolling stock used to establish the Sacramento Valley Railroad had to be imported by ship. Even with its own railroads, California remained as isolated as ever from the eastern United States and the rest of the world. Politicians, investors, and engineers such as Theodore Judah, Charles Crocker, Collis Huntington, and California governor Leland Stanford understood only too well that what the state needed was not so much its own internal railroad network as a reliable rail link to the Midwest and big cities in the East. In 1861, while the opening battles of the Civil War were being fought nearly 3,000 miles away in Virginia, these men and others formed the Central Pacific Railroad Company. Its aim was to build a railroad through the seemingly impenetrable wall of the Sierras and then push eastward toward the Rockies and the Mississippi. Congress passed bills encouraging the project, but until the war ended, the work went slowly. The first 50 miles of Central Pacific track, completed in 1864, cost $2.25 million and nearly bankrupted the company.

Completing the nation's first transcontinental railway would have to wait until after the Civil War, by which time it had become a joint project. From 1865 until the spring of 1869, the Central Pacific continued to blast and tunnel its way across the Sierras and the Great Basin while the Union Pacific pushed steadily westward across the Great Plains and the Rockies. The efforts of both companies were given a substantial boost by enormous federal subsidies—as much as $48,000 per mile of track.

The construction of the Central Pacific part of the line was made possible by up to ten thousand mostly Chinese laborers, who alternately sweated or froze while laying miles of track each day. Admiring the dedication and seemingly boundless energies of Chinese workers, Charles Crocker tended to favor them when hiring construction laborers. It is likely a myth, however, that most Chinese laborers were brought from across the Pacific for the sole purpose of building the railroad. In fact, many Chinese railroad laborers had originally come to California during the gold rush to prospect for gold or to take various jobs in the mining camps. Like so many others, they had gone bust and were forced to take whatever work they could find. For many the last resort proved to be a temporary job as a railroad construction laborer.

Though many sought them, these jobs were not particularly desirable. The work of lugging heavy steel rails, driving spikes, and breaking stone for bridges and embankments was extremely hard, and the pay amounted to little more than a dollar a day. Railroad construction work could also be very dangerous, and hundreds were killed while tunneling or blasting away stubborn ledges.

The last rail of the transcontinental line was put in place at Provo, Utah, on May 10, 1869, amid much ceremony. Appropriately, the final spike was made of pure California gold. California finally had its rail link to the rest of the nation. Easterners who might never have considered a trip to California now decided they'd like to pay the golden state a visit. Many who came never left, becoming part of a mighty tide of immigration that has still not ebbed to this day.

Early transcontinental rail journeys were adventurous affairs, subjecting travelers to discomforts and hazards such as blazing deserts, freezing mountain passes, hostile Indians, bandits, herds of buffalo, washed-out bridges, and broken-down locomotives. A trip from New York City to San Francisco could take several weeks. One daring tourist who took an early railroad excursion across the continent was novelist Robert Louis Stevenson, who would later write about the experience. Stevenson ended up in Monterey, where he stayed for a while as a guest at the Point Pinos Lighthouse. Interestingly enough, Stevenson was the scion of a famed family of lighthouse engineers in Great Britain, so he and the keeper must have had a lot to talk about. Some believe Stevenson wrote much of his classic novel *Treasure Island* while staying in Monterey.

Although less well known than construction of the Transcontinental Railroad, the establishment of lighthouses up and down the Pacific coast may have done even more to solidify California's ties to the rest of the civilized world. Alarmed by tragic and costly wrecks that took the lives of hundreds of gold rush miners and sometimes claimed their gold as well, the U.S. Congress funded a major lighthouse-construction effort along the West Coast. The first western lighthouse was completed at Alcatraz Island near San Francisco in 1854. Other lighthouses were soon brightening the rocks of Point Pinos, Point Loma near San Diego, Point Conception west of Santa Barbara, Point Bonita beside the Golden Gate, and at several other key points along the California coast. The guiding beacons of these bright sentinels saved many ships and lives and in this way contributed as much as or more than the railroads to the California we know today.

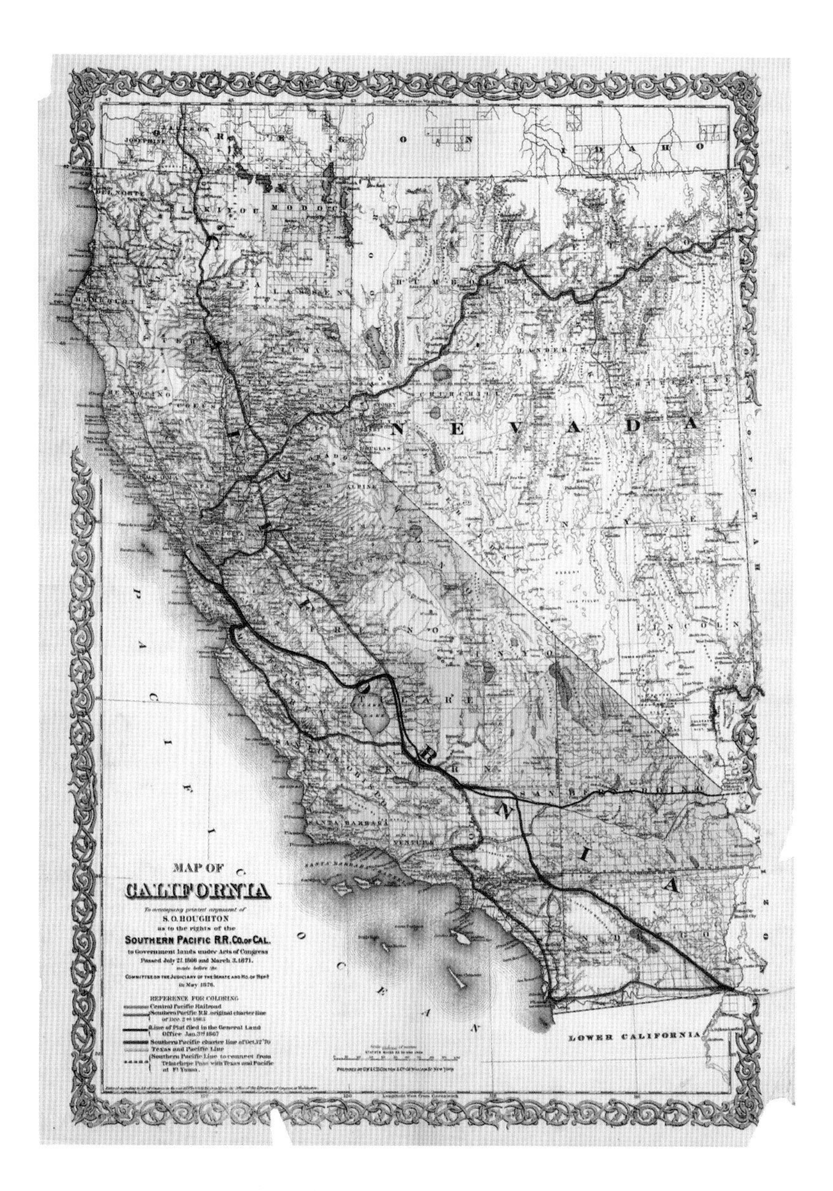

Map of California to accompany printed agreement of S. O. Houghton as to the rights of the Southern Pacific R.R. Co. of Cal. to government lands under Acts of Congress passed July 27, 1866, and March 3, 1871, made before the committee of the judiciary of the Senate and Ho. of Reps. in May 1876.

The federal government promoted construction not just of the Transcontinental Railroad but of railway lines throughout the West. One way this was done was by granting federal lands to railroad companies as a subsidy. The land was then sold to farmers and ranchers, who became dependent on the railroads to ship their cattle and produce to market. The map shown here accompanied copies of an 1876 land-transfer agreement benefiting California's Southern Pacific Railroad. The dark lines mark the routes of rail lines existing at that time. The route of the Transcontinental Railroad can be seen crossing Nevada and passing through the Sierras at Donner Pass.

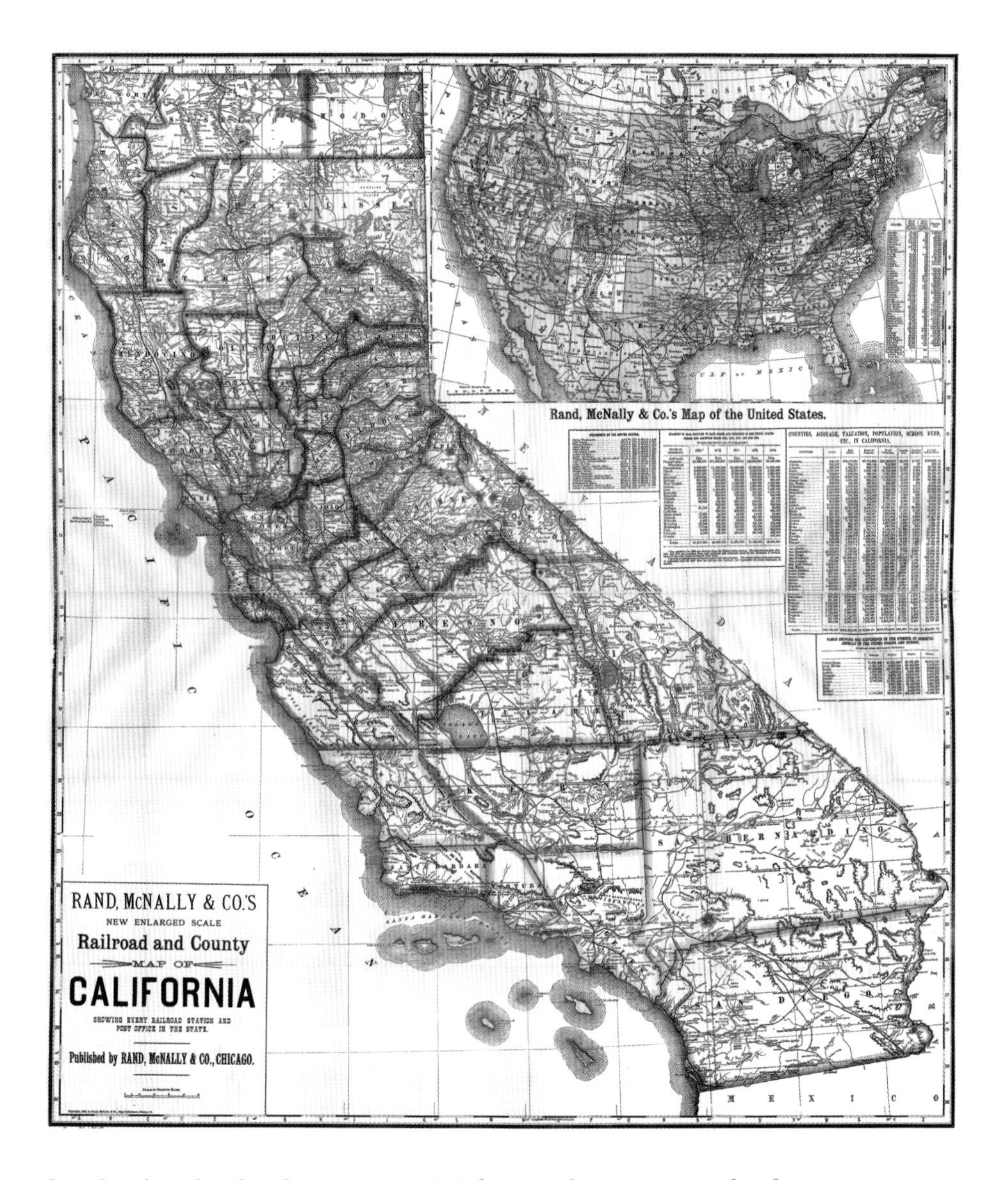

New enlarged scale railroad and county map of California showing every railroad station and post office in the state (1883).

As a network of tracks spread across the state, linking nearly every town of any size, trains became the prime mode of transportation in California. People could now travel between towns and cities with relative ease, and products could be shipped to market much faster. Trains carried the mail, which was often sorted en route in special railcars. Usually, the tracks were accompanied by telegraph lines so that news reached otherwise isolated communities as fast as a key could tap out code. This 1883 Rand McNally map shows existing counties along with every railway station and post office in the state.

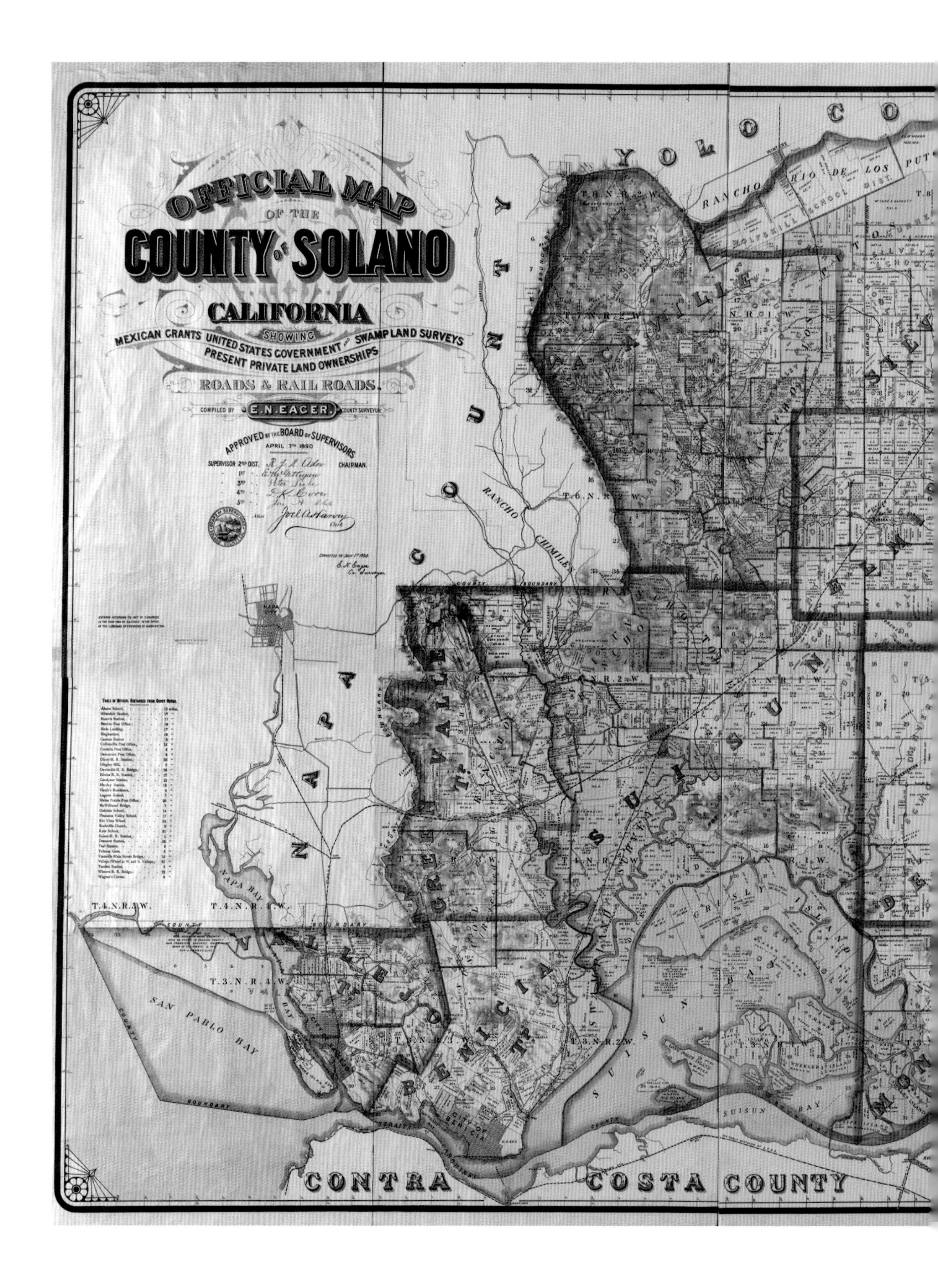
OFFICIAL MAP
OF THE
COUNTY OF SOLANO
CALIFORNIA
SHOWING
MEXICAN GRANTS UNITED STATES GOVERNMENT AND SWAMP LAND SURVEYS
PRESENT PRIVATE LAND OWNERSHIPS
ROADS & RAIL ROADS.
COMPILED BY E. N. EAGER. COUNTY SURVEYOR
APPROVED BY THE BOARD OF SUPERVISORS
APRIL 7TH 1890
SUPERVISOR 2ND DIST. CHAIRMAN.
YOLO CO
NAPA
CONTRA COSTA COUNTY
SAN PABLO BAY
NAPA BAY
SUISUN BAY
RANCHO CHIMILES
CITY OF BENICIA
T.4.N.R.3.W.
T.3.N.R.4.W.
T.6.N.R.1.W.

Official map of the County of Solano, California: showing Mexican grants, United States government and swamp land surveys, present private land ownerships, roads and railroads, Compiled by E. N. Eager, County Surveyor; approved by the Board of Supervisors (1890).

An 1890 map of Solano County shows roads, railroads, Mexican land grants, federal swampland surveys, and private holdings. Located about halfway between San Francisco and Sacramento, Solano County was positioned at the intersection of early transportation thoroughfares. The tracks of the Transcontinental Railroad traversed the county, and steamboats passed through it on their way to the state capital. Today Interstate 80 carries heavy traffic through the heart of the county. Although still partly rural, Solano County now has a population of about four hundred thousand.

MAP OF CALIFORNIA ROADS FOR CYCLERS
THE 3 LEADERS
THE COLUMBUS BUGGY CO'S
THE SPALDING
THE CREDENA
BICYCLES
COLUMBUS BUGGY CO
GENERAL AGENTS
213-215 MARKET ST SAN FRANCISCO
362-366 TWELFTH ST OAKLAND
N.B.-WRITE FOR CATALOGUES. WE MAKE A SPECIALTY OF WHOLESALING WHEELS.
THE PARK CYCLERY
LARGE RIDING SCHOOL
RENTING AND REPAIRING.
TERMINUS OF THE McALLISTER, POWELL AND GEARY ST. CAR LINES
A. TRUEWORTHY, Manager.
W.R. EVANS, STATE AGENT 1021 MARKET ST. S.F.
HELICAL TUBE PRIMIERS
Strongest & Lightest
PREMIER CYCLE Co
INCORPORATED San Francisco
Depot, Salesroom & Cyclery, 312-314 BAKER ST. East Entrance Panhandle of Park
RIDE
LOVELL DIAMOND
BICYCLES
GOLDEN WEST, SAN FRANCISCO SPECIAL, CAL. SPECIAL DIAMOND, 1896 WHEELS.
BIKE SPECIALTIES
SEND FOR CATALOGUE. L.S. MARTEL & Co 411½ CAL. ST. SAN FRANCISCO CAL.
AGENTS WANTED
THE BEST
SI-KO-LENE
LUBRICANT ON EARTH
SI-KO-LENE
FOR CYCLE CHAINS... TRY IT!
JOE E. SHEARER INVENTOR OF SI-KO-LENE
EXPLANATORY
RAILROADS
RAILROADS PROPOSED
BICYCLE ROADS
Condition
G....... GOOD
F....... FAIR
P....... POOR
V.P..... VERY POOR
Grade
L..... LEVEL
R..... ROLLING
H..... HILLY
M..... MOUNTAINOUS
Published and Copyrighted by GEO. W. BLUM San Francisco Cal.
THOS. E. KENT
FOR BICYCLE CLOTHING
597 MISSION ST. San Francisco
THE MARCH
THE SPEEDY PINK and BLUE
The REINFORCED FELLOES
The MARCH-DAVIS CYCLE CO.
J.G. FRENCH, PACIFIC COAST MANAGER
W.B. FAWCETT RETAIL MANAGER
739 MARKET ST San Francisco Cal.
PACIFIC OCEAN
SAN FRANCISCO
SACRAMENTO
STOCKTON
MODESTO
MERCED
TULARE LAKE
MONTEREY BAY
CARMEL BAY
SAN SIMEON BAY
CAYUCOS BAY
SANTA BARBARA CHANNEL
SCALE OF MILES
"NATIONAL" AND "DEERE"
BICYCLES
DEERE IMPLEMENT Co.
THE WHEELERY
TRIBUNE BICYCLES
C.M. CHRISTOFFER SOLE AGENT.
419 STANYAN ST. - SAN FRANCISCO.
RENTING, REPAIRING AND RIDING SCHOOL
TYLER BEACH PROPRIETOR
GEO. H. BEACH MANAGER
ST JAMES HOTEL
SAN JOSE, CAL.
OFFICIAL LEAGUE HOTEL.
REGULAR RATES, $2.00 to $3.00 Per Day
RATES TO CYCLERS, $2.00 Day
"UNION CRACKAJACK"
CRACKAJACKS RIDE UNIONS.
HAVE YOU SEEN THE BARNES-SPECIAL
HOOKER & Co. 16 & 18 DRUMM ST. SAN FRANCISCO.
This Map Engraved by the CALIFORNIA PHOTO-ENG. CO. Wasp Bld'g 513 Market St. S.F.
IF YOU RIDE WHY NOT RIDE THE BEST
THE VICTOR BICYCLE
FOR 16 YEARS THE MOST RELIABLE OF ALL BICYCLES.
OVERMAN WHEEL Co. MAKERS
OAKLAND LOS ANGELES
309 LARKIN ST. SAN FRANCISCO
COOPER CYCLE CO.
CITY AGENTS.
RENTING, REPAIRING AND RIDING SCHOOL.
THEY RIDE WHITE FLYERS
BARNES' SPECIAL
COR. BUSH AND DEVISADERO STS. SAN FRANCISCO, CAL.
O.S. Potter & Co.
Jobbers in High and Medium Grade Wheels Bicycle Clothing and Sundries Marion Wood Rims Clincher Hose Pipe Tires for '96.
Pacific Coast agents for
HALLADAY, WOLFF Sylph & Overland BICYCLES
511 MARKET ST. THREE DOORS ABOVE FIRST ST. SAN FRANCISCO
-DESIGNED & DRAWN BY GEO. W. BLUM-
-EDWARD DENNY & CO. 207 MONTGOMERY ST. SOLE AGENTS-

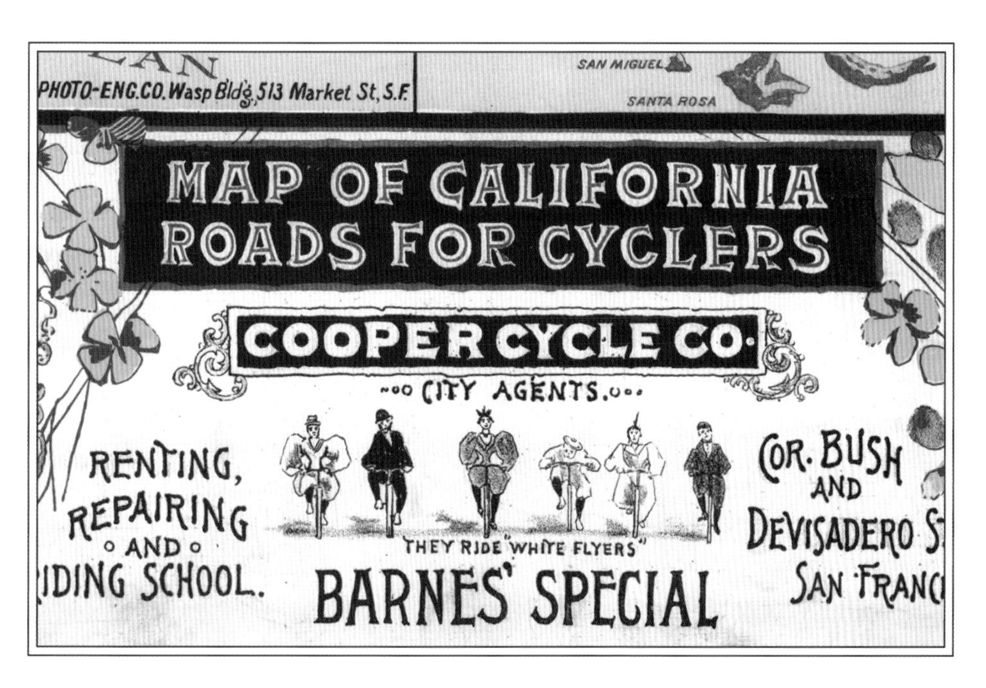

Map of California roads for cyclers (1895).

Bicycles, more or less as we know them today, were introduced during the late nineteenth century, and they quickly became a popular, even faddish, means of travel. Manufacturers across the country competed for the business of a bicycle-crazed public. Advertisements crowd the border of this 1895 map of California roads recommended for "cyclers." Famed for their leadership of so many trends, Californians were apparently out front on this one as well. Perhaps that should come as no surprise, since the state's weather and scenery make it especially attractive to cyclists.

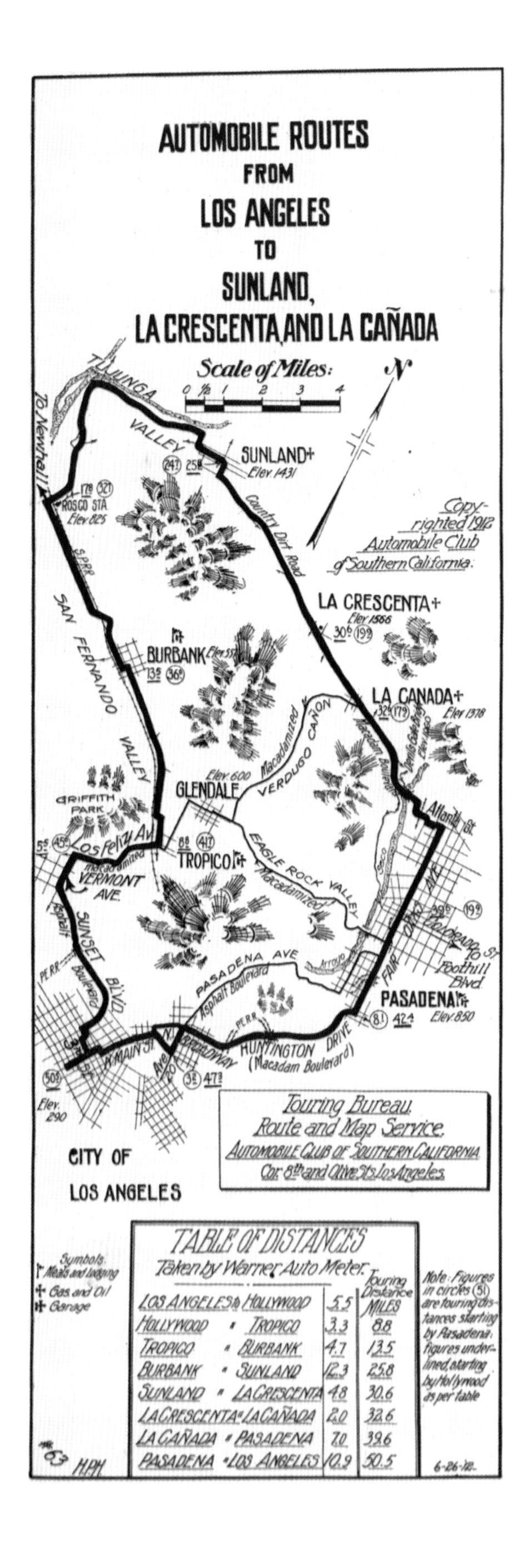

TABLE OF DISTANCES
Taken by Warner Auto Meter.

			Touring Distance Miles
LOS ANGELES to	HOLLYWOOD	5.5	
HOLLYWOOD "	TROPICO	3.3	8.8
TROPICO "	BURBANK	4.7	13.5
BURBANK "	SUNLAND	12.3	25.8
SUNLAND "	LA CRESCENTA	4.8	30.6
LA CRESCENTA "	LA CAÑADA	2.0	32.6
LA CAÑADA "	PASADENA	7.0	39.6
PASADENA "	LOS ANGELES	10.9	50.5

Golden Cities

San Francisco is blessed by what may be the largest and best harbor in the world. The city's harbor is in effect the whole of the 70-mile-long and 20-mile-wide San Francisco Bay. It may be that the presence of such a maritime resource all but ensured the city's preeminence among western communities, but the discovery of gold in the Sierra foothills only a few dozen miles to the east made it inevitable. During the time of the padres, San Francisco was a village consisting of a few families whose homes clustered around a mission church and a presidio. By the time the gold rush began in earnest in 1849, the city could still claim no more than a thousand residents. Within five years, however, its population had ballooned to twenty-five thousand.

During the gold rush era, San Francisco was essentially a larger and only slightly more settled version of a rough-and-ready mining camp. There were the same reeking saloons, the same bawdy houses, the same lawlessness,

Automobile routes from Los Angeles to Sunland, La Crescenta, and La Cañada, Touring bureau. Route and map service. Automobile Club of Southern California (1912).

A powerful new transportation phenomenon—the "horseless carriage"—arrived in California during the early twentieth century, and it would play a key role in the development of sprawling cities like Los Angeles, San Diego, and Bakersfield. Automobiles changed the way most Americans viewed their world, but no one would take to them with quite the same passion as Californians. Although too expensive for most families at first, automobiles were becoming more and more common by the time this auto tour guide of the Los Angeles area was published in 1912. It sketches out a scenic route from Los Angeles through the San Fernando Valley and back by way of Pasadena. An all-day excursion a century ago, the same trip can be made today in a couple of hours depending on the freeway traffic.

and the same panhandling prospectors looking for a grubstake. The difference was that many who came to San Francisco during the late 1840s and early 1850s did not pack up and leave when the gold strikes played out. Those who wanted to stay or were forced to remain because of their financial circumstances discovered that there were jobs to be had in enterprises other than mining and money to be made in ways other than panning for gold. Some opened hardware stores, clothing shops, and bakeries. Others sold candy, eggs, beef, or butter. Still others became policemen or professional firefighters, and a few, such as early San Francisco mayor John Geary, threw their hats into the political arena.

People liked San Francisco. They liked the weather, despite Mark Twain's quip that "the coldest winter I ever spent was a summer in San Francisco." They liked the views of the mountains and the bay. They liked the freewheeling atmosphere of the place that gave one the feeling that anything could happen. Many found that they could get rich in San Francisco and do it without finding gold. Levi Strauss managed to pile up an enormous fortune by manufacturing work pants held together by rivets—nowadays we usually call them "blue jeans."

Following the gold rush, San Francisco quickly became the Far West's dominant commercial hub and its only true urban center. By 1870 the population of San Francisco had surpassed 250,000, more than 250 times the city's size twenty years earlier, a remarkable growth spurt seldom exceeded in modern times. The city grew at a much slower rate later in the nineteenth century and throughout the twentieth century, though its success as a business center remained undiminished. In recent decades San Francisco's fortunes have risen still further as the dash to develop ever more capable and profitable computers has touched off a new sort of gold rush in the nearby Silicon Valley.

Unlike San Francisco, Los Angeles changed very little during the gold rush years. Like San Francisco, the City of Angels began the 1850s with fewer than a thousand residents. By 1870 it had grown significantly but could boast a population of only about five thousand, which made it a mere village by San Francisco standards. What held Los Angeles back was commerce, or the lack of it. There had been little gold in the hills east of the city to fuel a boom. As it turned out, however, there was gold of a different kind lying practically beneath the city's streets. In 1892 a man named Edward Doheny struck oil near the present-day site of Dodger Stadium.

Within a few years the Los Angeles area had become one of the world's leading oil producers, and as more and more of this black gold was pumped out of the ground and other local industries grew, more and more Americans decided to move to Los

Angeles. Some were attracted by the beaches and orange groves, others by the glitz of the movie industry that took root in Hollywood about 1910 and has flourished in the Southern California sun ever since. By 1930 the city's population was more than a million, making Los Angeles far larger than its rival to the north. The Los Angeles area has a population of about 14 million today.

Of course, Los Angeles nowadays is not just one city. It is many cities. Instead of having a more-or-less-fixed urban core like San Francisco, the city and its many suburbs are spread out all over the Los Angeles basin. Los Angeles has no fewer than seventy major suburbs, which lumped together create one giant urbanized area nearly 100 miles long and more than 50 miles across. Because the city is so extensive, Angelinos—that's how the citizens of Los Angeles refer to themselves—rely very heavily on automobiles. Given the sprawl and the all-too-often jammed-up condition of the freeways, it may require a drive of an hour and a half or more for one Los Angeles–area resident to visit another or to take in a museum or a movie downtown.

The strike-it-rich mentality that helped create the San Francisco or Los Angeles we know today also fueled the growth of other California cities. However, the boom times these cities enjoyed were not always based on oil, gold, or other precious metals. Eureka is a quaint and quiet Victorian-style community today, but a century ago its ambience was much like that of a gold rush mining camp. The substance that drove its fast-moving economic engine was not a mineral at all, but rather the rich red wood of the coastal sequoia. Other California cities and smaller communities would have their own boom times based on whatever products their citizens could use to generate wealth. For Palo Alto and Berkeley, that product was a superior education. For Napa and Sonoma it was wine. For San Diego it was a homeport for the U.S. Navy. For Fresno and Modesto it was access to some of the world's greatest natural wonders. And more recently, for San Jose it was silicon chips and computer technology.

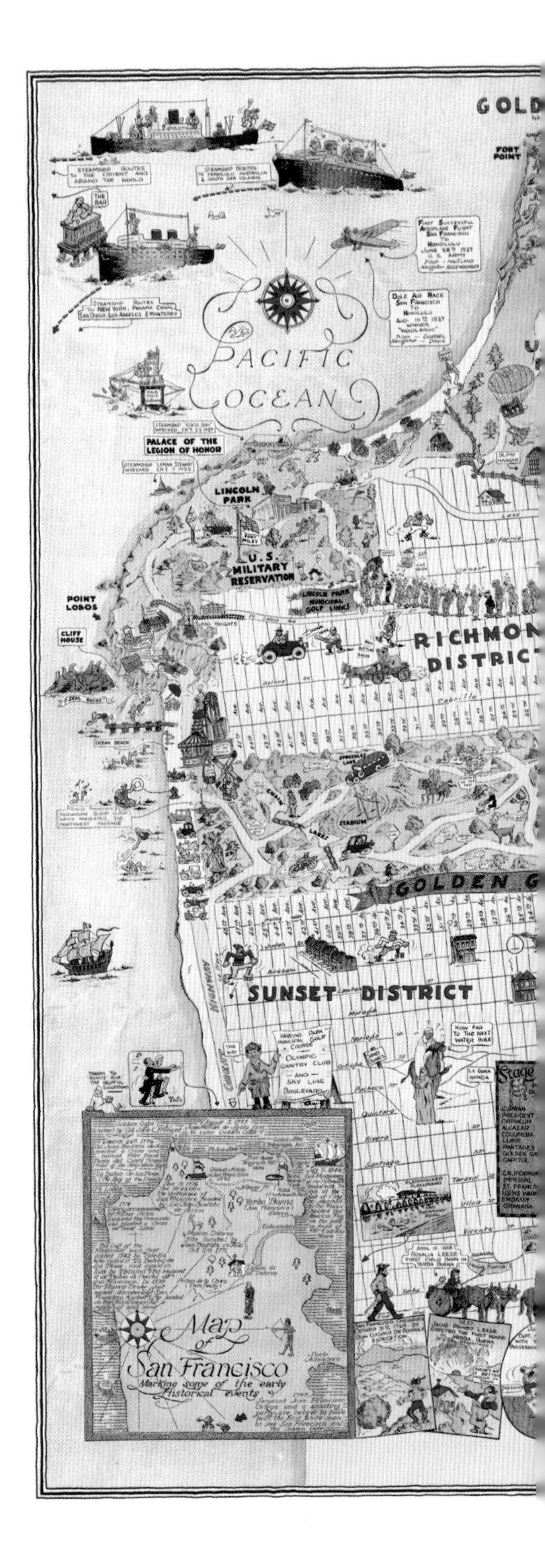

Map of San Francisco, showing principal streets and places of interest (1927).

A small town when the United States took possession of California during the late 1840s, San Francisco was transformed almost overnight by the gold rush. Its population grew from less than a thousand to nearly a quarter million in about twenty years. For decades the largest city in the West, San Francisco would eventually be eclipsed by Los Angeles—in population at least—but it remains to this day one of the world's great cities. A delightfully illustrated tourist map makes it clear the City by the Bay was just as lively a place during the Roaring Twenties as it is today. An interesting characteristic of this map is that it was published prior to construction of the Golden Gate Bridge, which was completed in 1937.

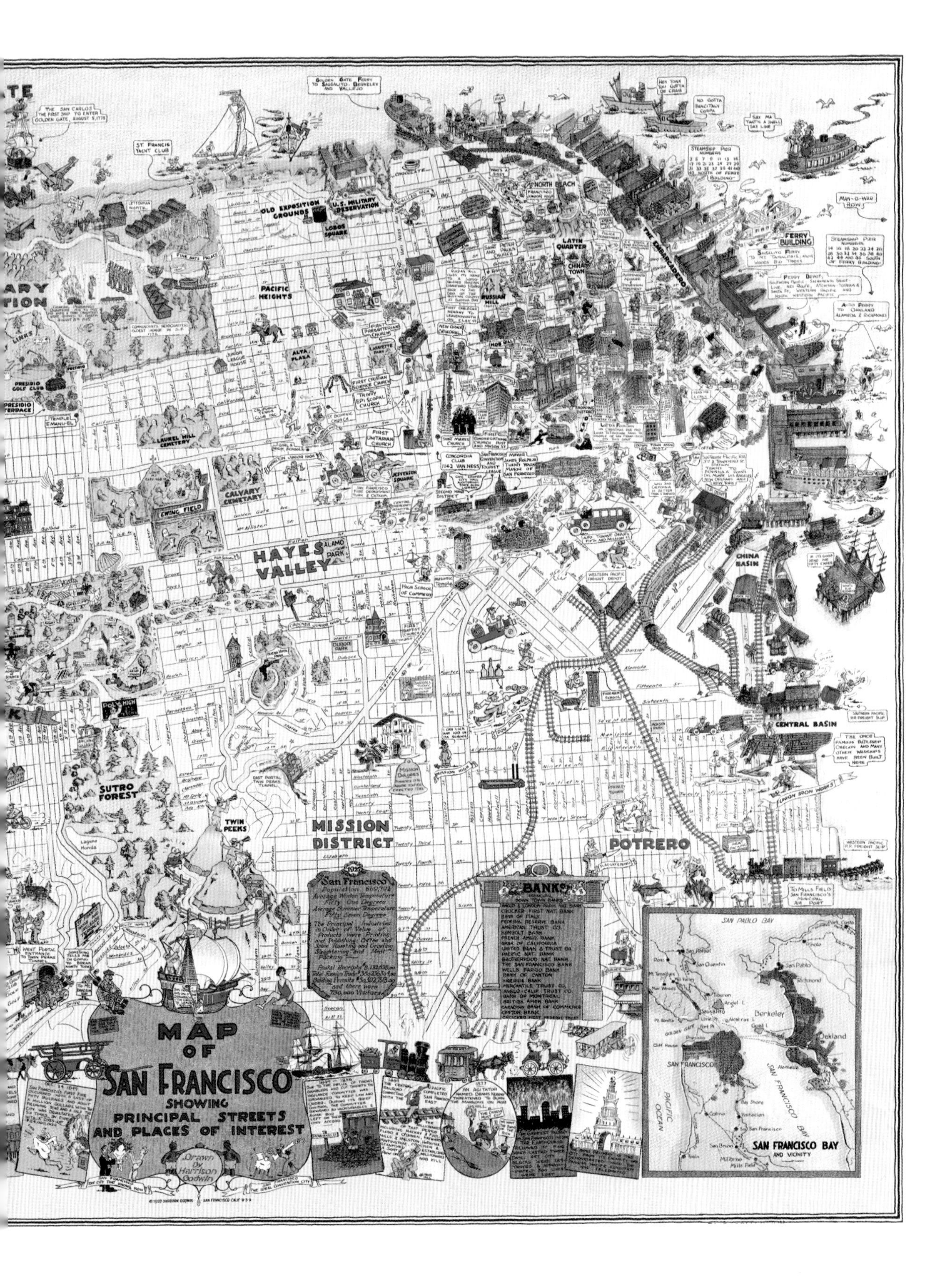
MAP OF SAN FRANCISCO
SHOWING PRINCIPAL STREETS AND PLACES OF INTEREST
Drawn By Harrison Godwin
SAN FRANCISCO THE CITY THAT KNOWS HOW
SAN FRANCISCO THE IDEAL CONVENTION CITY
© 1927 HARRISON GODWIN · SAN FRANCISCO CALIF U S A
THE SAN CARLOS THE FIRST SHIP TO ENTER GOLDEN GATE, AUGUST 5, 1775
GOLDEN GATE FERRY TO SAUSALITO, BERKELEY AND VALLEJO
ST FRANCIS YACHT CLUB
HEY TONY YOU GOTTA DA CRAB
NO GOTTA BANCITTALY CORPA
SAY MA THAT'S A SMALL SAILBOAT
STEAMSHIP PIER NUMBERS
NORTH BEACH
OLD EXPOSITION GROUNDS
U.S. MILITARY RESERVATION
LOBOS SQUARE
MAN-O-WAR ROW
FERRY BUILDING
THE EMBARCADERO
LATIN QUARTER
CHINA TOWN
RUSSIAN HILL
PACIFIC HEIGHTS
NOB HILL
ALTA PLAZA
PRESIDIO GOLF CLUB
PRESIDIO TERRACE
TEMPLE EMANU-EL
LAUREL HILL CEMETERY
CALVARY CEMETERY
JEFFERSON SQUARE
EWING FIELD
HAYES VALLEY
ALAMO PARK
CONCORDIA CLUB 1142 VAN NESS
CHINA BASIN
CENTRAL BASIN
UNION IRON WORKS
SUTRO FOREST
TWIN PEEKS
MISSION DISTRICT
POTRERO
1925
San Francisco
Population 669,792
Average Winter Temperature Fifty One Degrees
Average Summer Temperature Fifty Seven Degrees
The Principal Industries in Order of Value of Products were Printing and Publishing; Coffee and Spice Roasting and Grinding; Slaughtering and Meat Packing
Postal Receipts $8,132,898.00
Total Foreign Trade $380,236,304.00
Building Permits $50,392,793.00
and there were 730,000 Visitors
BANKS
PRINCIPAL DOWN TOWN BANKS
CROCKER FIRST NAT. BANK
BANK OF ITALY
AMERICAN TRUST CO.
HOMBOLT BANK
FRENCH AMER. BANK
BANK OF CALIFORNIA
UNITED BANK & TRUST CO.
PACIFIC NAT. BANK
BROTHERHOOD NAT. BANK
THE SAN FRANCISCO BANK
WELLS FARGO BANK
BANK OF CANTON
HIBERNIA BANK
MERCANTILE TRUST CO.
ANGLO-CALIF. TRUST CO.
BANK OF MONTREAL
BRITISH AMER. BANK
CANADIAN BANK OF COMMERCE
CANTON BANK
1877 AN AGITATOR NAMED DENNIS KEARNY THREATENED TO BURN THE MANSIONS ON NOB HILL
1915
SAN PABLO BAY
San Rafael
San Quentin
San Pablo
Sausalito
Alcatraz I.
Angel I.
Berkeley
Oakland
Alameda
SAN FRANCISCO
PACIFIC OCEAN
San Bruno
Millbrae
SAN FRANCISCO BAY
SAN FRANCISCO BAY AND VICINITY

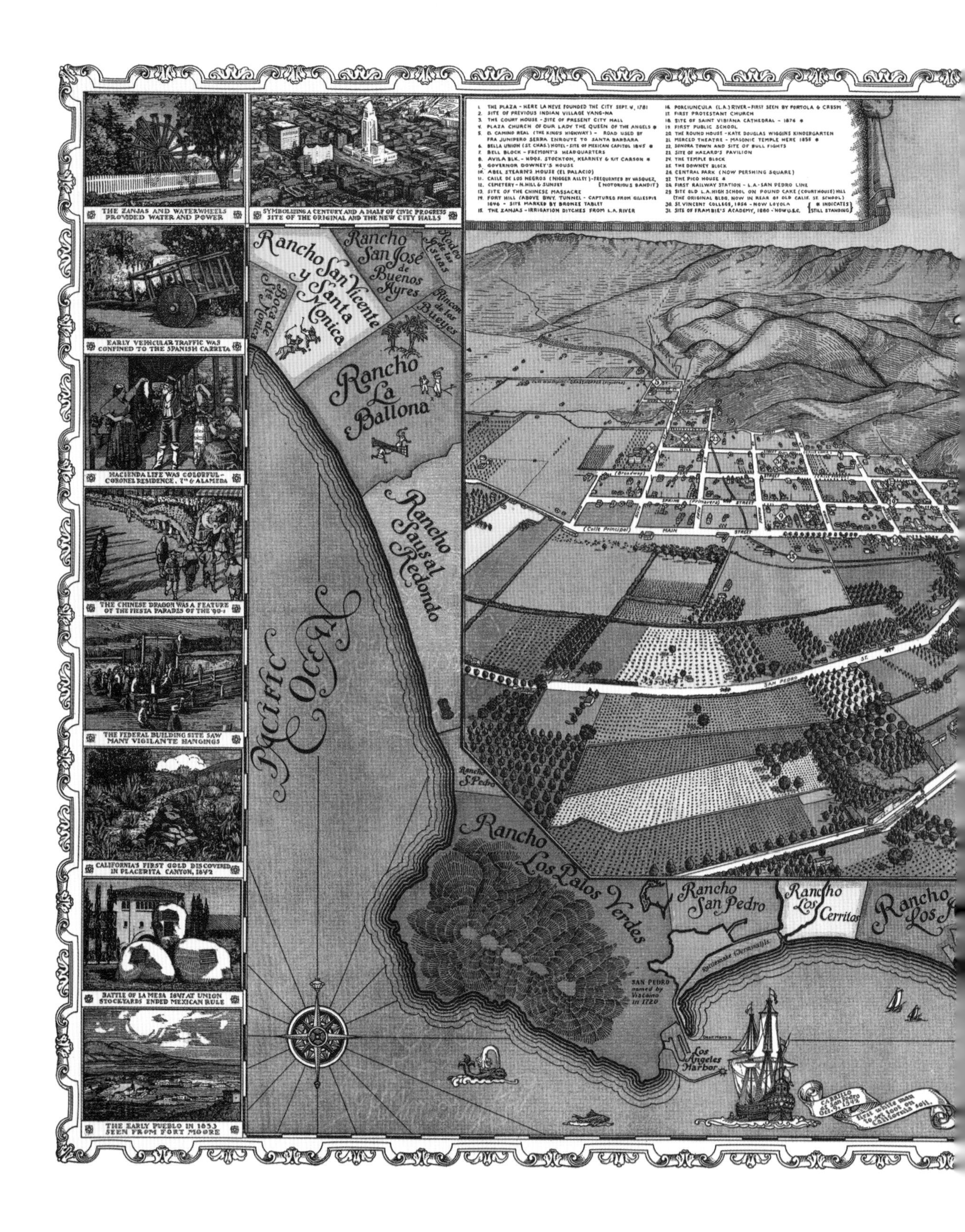
THE ZANJAS AND WATERWHEELS PROVIDED WATER AND POWER
SYMBOLIZING A CENTURY AND A HALF OF CIVIC PROGRESS SITE OF THE ORIGINAL AND THE NEW CITY HALLS
EARLY VEHICULAR TRAFFIC WAS CONFINED TO THE SPANISH CARRETA
HACIENDA LIFE WAS COLORFUL - CORONEL RESIDENCE, 7TH & ALAMEDA
THE CHINESE DRAGON WAS A FEATURE OF THE FIESTA PARADES OF THE '90s
THE FEDERAL BUILDING SITE SAW MANY VIGILANTE HANGINGS
CALIFORNIA'S FIRST GOLD DISCOVERED IN PLACERITA CANYON, 1842
BATTLE OF LA MESA 1847 AT UNION STOCKYARDS ENDED MEXICAN RULE
THE EARLY PUEBLO IN 1853 SEEN FROM FORT MOORE
1. THE PLAZA - HERE LA NEVE FOUNDED THE CITY SEPT. 4, 1781
2. SITE OF PREVIOUS INDIAN VILLAGE YANG-NA
3. THE COURT HOUSE - SITE OF PRESENT CITY HALL
4. PLAZA CHURCH OF OUR LADY THE QUEEN OF THE ANGELS *
5. EL CAMINO REAL (THE KING'S HIGHWAY) - ROAD USED BY FRA JUNIPERO SERRA ENROUTE TO SANTA BARBARA
6. BELLA UNION (ST. CHAS.) HOTEL - SITE OF MEXICAN CAPITOL 1845 *
7. BELL BLOCK - FREMONT'S HEADQUARTERS
8. AVILA BLK. - HDQS. STOCKTON, KEARNEY & KIT CARSON *
9. GOVERNOR DOWNEY'S HOUSE
10. ABEL STEARN'S HOUSE (EL PALACIO)
11. CALLE DE LOS NEGROS (NIGGER ALLEY) - FREQUENTED BY VASQUEZ, (NOTORIOUS BANDIT)
12. CEMETERY - N. HILL & SUNSET
13. SITE OF THE CHINESE MASSACRE
14. FORT HILL (ABOVE BWY. TUNNEL - CAPTURED FROM GILLESPIE 1846 - SITE MARKED BY BRONZE TABLET
15. THE ZANJAS - IRRIGATION DITCHES FROM L.A. RIVER
16. PORCIUNCULA (L.A.) RIVER - FIRST SEEN BY PORTOLA & CRESPI
17. FIRST PROTESTANT CHURCH
18. SITE OF SAINT VIBIANA CATHEDRAL - 1876 *
19. FIRST PUBLIC SCHOOL
20. THE ROUND HOUSE - KATE DOUGLAS WIGGINS KINDERGARTEN
21. MERCED THEATRE - MASONIC TEMPLE HERE 1855 *
22. SONORA TOWN AND SITE OF BULL FIGHTS
23. SITE OF HAZARD'S PAVILION
24. THE TEMPLE BLOCK
25. THE DOWNEY BLOCK
26. CENTRAL PARK (NOW PERSHING SQUARE)
27. THE PICO HOUSE *
28. FIRST RAILWAY STATION - L.A.-SAN PEDRO LINE
29. SITE OLD L.A. HIGH SCHOOL ON POUND CAKE (COURTHOUSE) HILL (THE ORIGINAL BLDG. NOW IN REAR OF OLD CALIF. ST. SCHOOL)
30. ST. VINCENT COLLEGE, 1856 - NOW LOYOLA
31. SITE OF FRAMBIE'S ACADEMY, 1880 - NOW U.S.C.
* INDICATES STILL STANDING
Rancho San José de Buenos Ayres
Rancho San Vicente y Santa Monica
Boca de Sta Monica
Rincon de los Bueyes
Rancho La Ballona
Rancho Sausal Redondo
PACIFIC OCEAN
Rancho Los Palos Verdes
Rancho San Pedro
Rancho Los Cerritos
Rancho Los
SAN PEDRO named by Viscaino in 1720
Los Angeles Harbor
(Broadway)
(Calle Principal)
MAIN
STREET
SAN PEDRO

Los Angeles as it appeared in 1871. Gores, fecit.

An 1871 view pictures Los Angeles as a quiet agricultural town. The map's colorful illustrations highlight area attractions, so even way back then the City of Angels apparently drew its share of tourists. Many of them must have decided to stay. The California Gold Rush had little impact on Los Angeles, and by the 1870s it had a cathedral but a population of only about five thousand. Within a few decades, however, oil, oranges, and a burgeoning movie industry would alter the city's fortunes dramatically. By 1930 the city's population had passed the 1 million mark and was still growing at a record pace. Today Los Angeles and its extensive suburbs house about 14 million people.

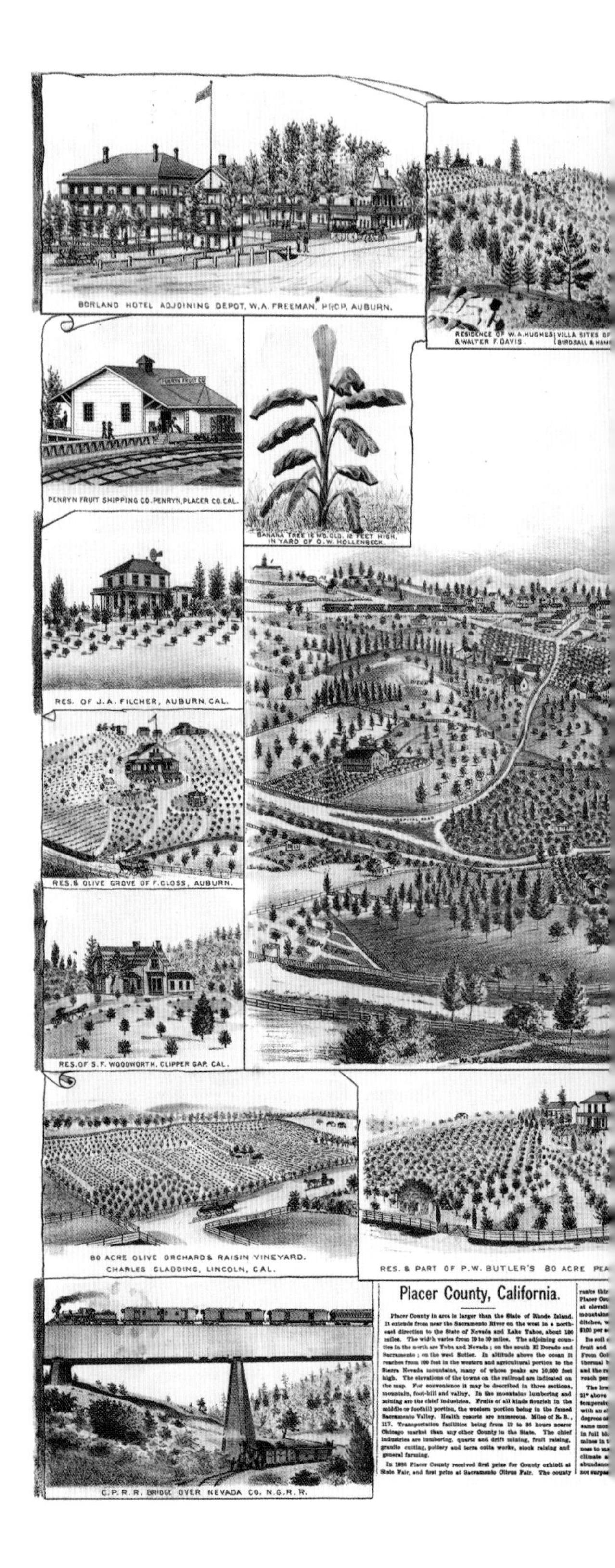

Auburn, Cal., C. P. Cook., del.; presented with the compliments of W. B. Lardner, real estate agent, att'y-at-law & notary public (1887).

Privately published by W. B. Lardner, who gave copies to his insurance and real estate clients, an 1880s view of Auburn depicts it as a handsome rural community, just the sort of place in which a family might want to buy property and settle down. Actually, the town got its start as an 1840s mining camp beset by mud, squalor, and all the other unattractive qualities of such places. Its original name was North Fork Dry Diggings. Today Auburn is a lovely foothills community of about twelve thousand that takes great pride in its gold rush heritage.

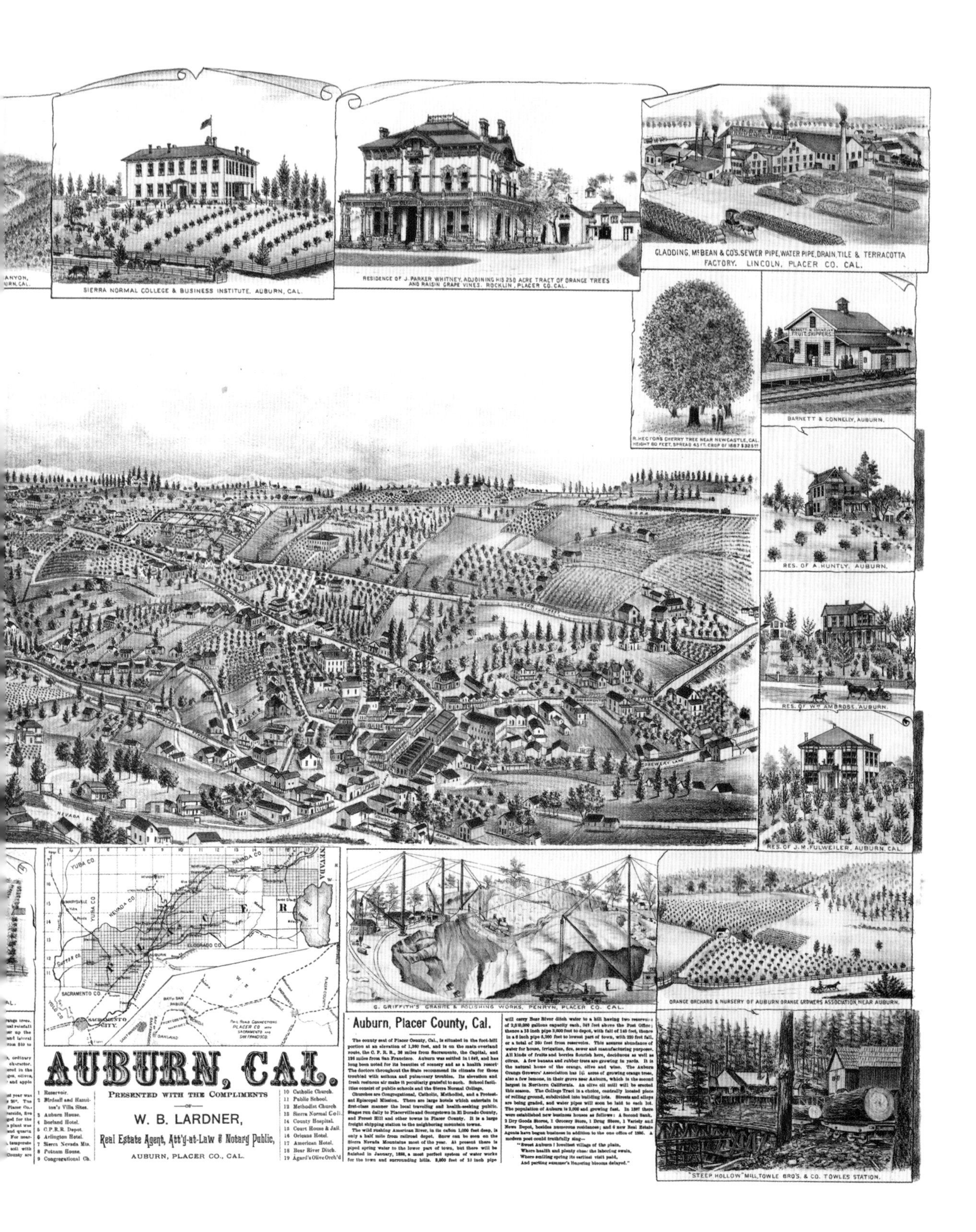
SIERRA NORMAL COLLEGE & BUSINESS INSTITUTE, AUBURN, CAL.
RESIDENCE OF J. PARKER WHITNEY, ADJOINING HIS 250 ACRE TRACT OF ORANGE TREES AND RAISIN GRAPE VINES. ROCKLIN, PLACER CO. CAL.
GLADDING, McBEAN & CO'S. SEWER PIPE, WATER PIPE, DRAIN, TILE & TERRACOTTA FACTORY. LINCOLN, PLACER CO. CAL.
BARNETT & CONNELLY, AUBURN.
RES. OF A. HUNTLY, AUBURN.
RES. OF Wm AMBROSE, AUBURN.
RES. OF J. M. FULWEILER, AUBURN, CAL.
ORANGE ORCHARD & NURSERY OF AUBURN ORANGE GROWERS ASSOCIATION, NEAR AUBURN.
G. GRIFFITH'S GRANITE & POLISHING WORKS, PENRYN, PLACER CO. CAL.
"STEEP HOLLOW" MILL, TOWLE BRO'S. & CO. TOWLES STATION.
NEVADA ST.
BREWERY LANE
YUBA CO.
NEVADA CO.
NEVADA
SUTTER CO.
SACRAMENTO CO.
ELDORADO CO.
AUBURN
SACRAMENTO CITY
OAKLAND
AUBURN, CAL.
1 Reservoir.
2 Birdsall and Hamilton's Villa Sites.
3 Auburn House.
4 Borland Hotel.
5 C.P.R.R. Depot.
6 Arlington Hotel.
7 Sierra Nevada Mts.
8 Putnam House.
9 Congregational Ch.
PRESENTED WITH THE COMPLIMENTS
—OF—
W. B. LARDNER,
Real Estate Agent, Att'y-at-Law & Notary Public,
AUBURN, PLACER CO., CAL.
10 Catholic Church.
11 Public School.
12 Methodist Church.
13 Sierra Normal Coll.
14 County Hospital.
15 Court House & Jail.
16 Orleans Hotel.
17 American Hotel.
18 Bear River Ditch.
19 Agard's Olive Orch'd
Auburn, Placer County, Cal.
The county seat of Placer County, Cal., is situated in the foot-hill portion at an elevation of 1,380 feet, and is on the main overland route, the C. P. R. R., 36 miles from Sacramento, the Capital, and 126 miles from San Francisco. Auburn was settled in 1849, and has long been noted for its beauties of scenery and as a health resort. The doctors throughout the State recommend its climate for those troubled with asthma and pulmonary troubles. Its elevation and fresh resinous air make it peculiarly grateful to such. School facilities consist of public schools and the Sierra Normal College.
Churches are Congregational, Catholic, Methodist, and a Protestant Episcopal Mission. There are large hotels which entertain in first-class manner the local traveling and health-seeking public. Stages run daily to Placerville and Georgetown in El Dorado County, and Forest Hill and other towns in Placer County. It is a large freight shipping station to the neighboring mountain towns.
The wild rushing American River, in its cañon 1,000 feet deep, is only a half mile from railroad depot. Snow can be seen on the Sierra Nevada Mountains most of the year. At present there is piped spring water to the lower part of town, but there will be finished in January, 1888, a most perfect system of water works for the town and surrounding hills. 3,800 feet of 11 inch pipe will carry Bear River ditch water to a hill having two reservoirs of 2,000,000 gallons capacity each, 245 feet above the Post Office; thence a 13 inch pipe 3,800 feet to depot, with fall of 140 feet, thence in a 6 inch pipe 6,300 feet to lowest part of town, with 220 feet fall, or a total of 360 feet from reservoirs. This assures abundance of water for house, irrigation, fire, sewer and manufacturing purposes. All kinds of fruits and berries flourish here, deciduous as well as citrus. A few banana and rubber trees are growing in yards. It is the natural home of the orange, olive and vine. The Auburn Orange Growers' Association has 16 acres of growing orange trees, also a few lemons, in their grove near Auburn, which is the second largest in Northern California. An olive oil mill will be erected this season. The College Tract is a choice, centrally located piece of rolling ground, subdivided into building lots. Streets and alleys are being graded, and water pipes will soon be laid to each lot. The population of Auburn is 2,000 and growing fast. In 1887 there were established new business houses as follows: A Second Bank, 3 Dry Goods Stores, 1 Grocery Store, 1 Drug Store, 1 Variety and News Depot, besides numerous residences; and 6 new Real Estate Agents have begun business in addition to the one office of 1886. A modern poet could truthfully sing—
"Sweet Auburn! loveliest village of the plain,
Where health and plenty cheer the laboring swain,
Where smiling spring its earliest visit paid,
And parting summer's lingering blooms delayed."

A. T. & S. F. DEPOT, AZUSA.

Bird's eye view of Azusa, Los Angeles Co. Cal.,
E. S. Moore, del. (1887).

In 1887 John Slauson and a group of allied developers laid out the town of Azusa, and this bird's-eye view created at about that time illustrates their handiwork. Azusa is located in western Los Angeles County at the foot of the San Gabriel Mountains near the entrance to the canyon carved out by the San Gabriel River. As Slauson's depiction makes clear, the town had a hotel, a railway station, and a network of streets lined by a few houses but not much else. Today Azusa is a suburban community of about forty-five thousand. Local legend holds that Azusa is a compaction of "A to Z in the USA." Actually, Azusa is derived from Asuksagna, the Indian name for the canyon entrance.

Bird's eye view of Coronado Beach, San Diego Bay and city of San Diego, Cal. in distance, sketch by E. S. Moore; Crocker & Co., lith., S.F. (188?).

Explorer Juan Rodríguez Cabrillo became the first European in Upper California when he sailed into San Diego Bay in 1542. However, a permanent settlement was not established here until 1769, when Father Junipero Serra selected San Diego as the site for the first of his missions. The many vessels going and coming in this 1880s maplike view points out San Diego's chief advantage for settlers—it has one of the finest harbors in the West. This made San Diego an ideal location for naval facilities, and these have long provided an important boost to the local economy. The city now ranks among the largest and most prosperous cities in the United States.

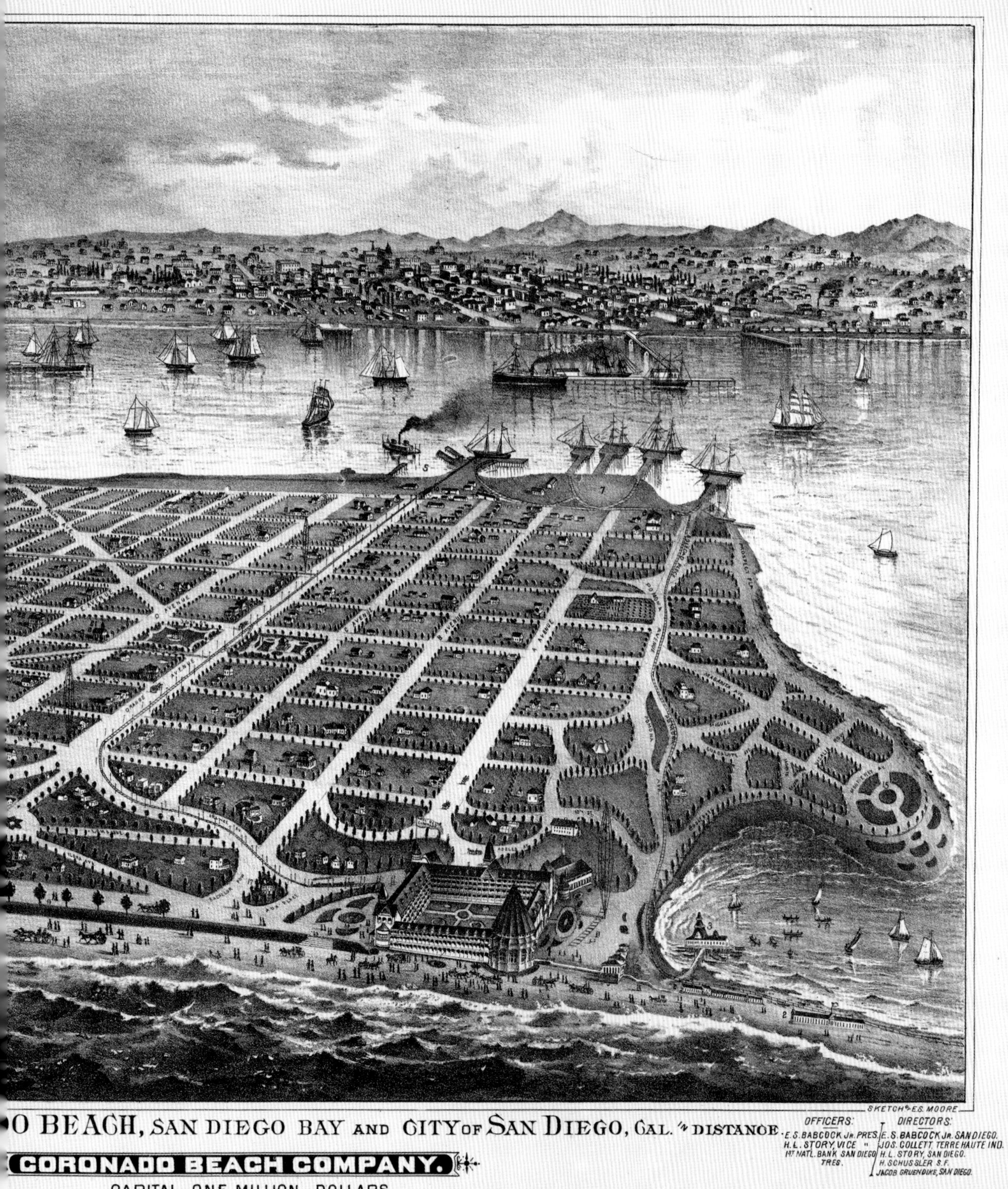

SKETCH & E.S. MOORE
O BEACH, SAN DIEGO BAY AND CITY OF SAN DIEGO, CAL. IN DISTANCE
CORONADO BEACH COMPANY.
CAPITAL-ONE MILLION DOLLARS.
OFFICERS:
E. S. BABCOCK JR. PRES.
H. L. STORY, VICE "
1ST NATL. BANK SAN DIEGO TRES.
DIRECTORS:
E. S. BABCOCK JR. SAN DIEGO.
JOS. COLLETT, TERRE HAUTE IND.
H. L. STORY, SAN DIEGO.
H. SCHUSSLER S.F.
JACOB GRUENDIKE, SAN DIEGO.

FRESNO HIGH SCHOOL
RESIDENCE OF Dr. D.H. TROWBRIDGE.
Mc DOUGALL BROS. ARCHITECTS. SAN FRANCISCO & FRESNO.
SUPT. PUBLIC CONSTRUCTION PLANS & ESTIMATES FURNISHED ON SHORT NOTICE
F.R. SMITH CONTRACTOR & BUILDER
ST. GEORGE VINEYARD
COURT HOUSE
THE FIRST NATIONAL BANK
SAN JOAQUIN ICE CO. AND CREAMERY.
SAN JOAQUIN ICE CO.
SEROPIAN BROTHERS PACKERS & SHIPPERS OF
J.B. INDERRIEDEN CO.
PACKERS of DRIED FRUITS AND RAISINS.
WORMSER & JONES
BARTON OPERA HOUSE
OIL LANDS & REAL ESTATE
MARKARIAN & CO.
RAISINS & FIGS
EVAPORATED FRUIT
SHEPHERD & TEAGUE
THE DORSEY-ROBINSON CO. GROCERS
CALWA WINERY AND DISTILLERY
CALIFORNIA WINE ASSOCIATION
FRESNO COUNTY ABSTRACT COMPANY
HIGH GRADE SWEET WINES AND GRAPE BRANDIES.
REEDLEY WINERY & DISTILLERY
MADARY'S PLANING MILL
DOORS, WINDOWS, BLINDS, SCREENS BOXES & TRAYS
M.R. MADARY.
FRESNO
CALIFORNIA
Copyrighted 1901 by L.W. KLEIN.
1901

Fresno, California (1901).

Insets featuring attractions and key government, business, and manufacturing facilities ring a 1901 view of Fresno. Interestingly, they do not emphasize the city's primarily agricultural roots and economic underpinnings. Founded during the 1870s, shortly after the first trains chugged through here, Fresno became the center of an agricultural empire based on irrigation and year-round cultivation of rich Central Valley soils. By the mid-twentieth century Fresno County had become the most productive farm county in America. Today Fresno struggles to keep up with its burgeoning population, which has surpassed five hundred thousand.

Bakersfield, Kern County, California (1901). There was so much groundwater beneath the San Joaquin Valley that it once pooled on the surface, creating shallow, mosquito-infested lakes. Pioneers built log cabins on an island in these marshes about 1860, and this small rustic community eventually became the city of Bakersfield. The aerial view shown here depicts the city as it appeared in 1901, when it had only about 5,000 residents. Today it boasts a population of more than 340,000, making it the eleventh largest city in California. Like that of Los Angeles, Bakersfield's growth was given a boost by oil.

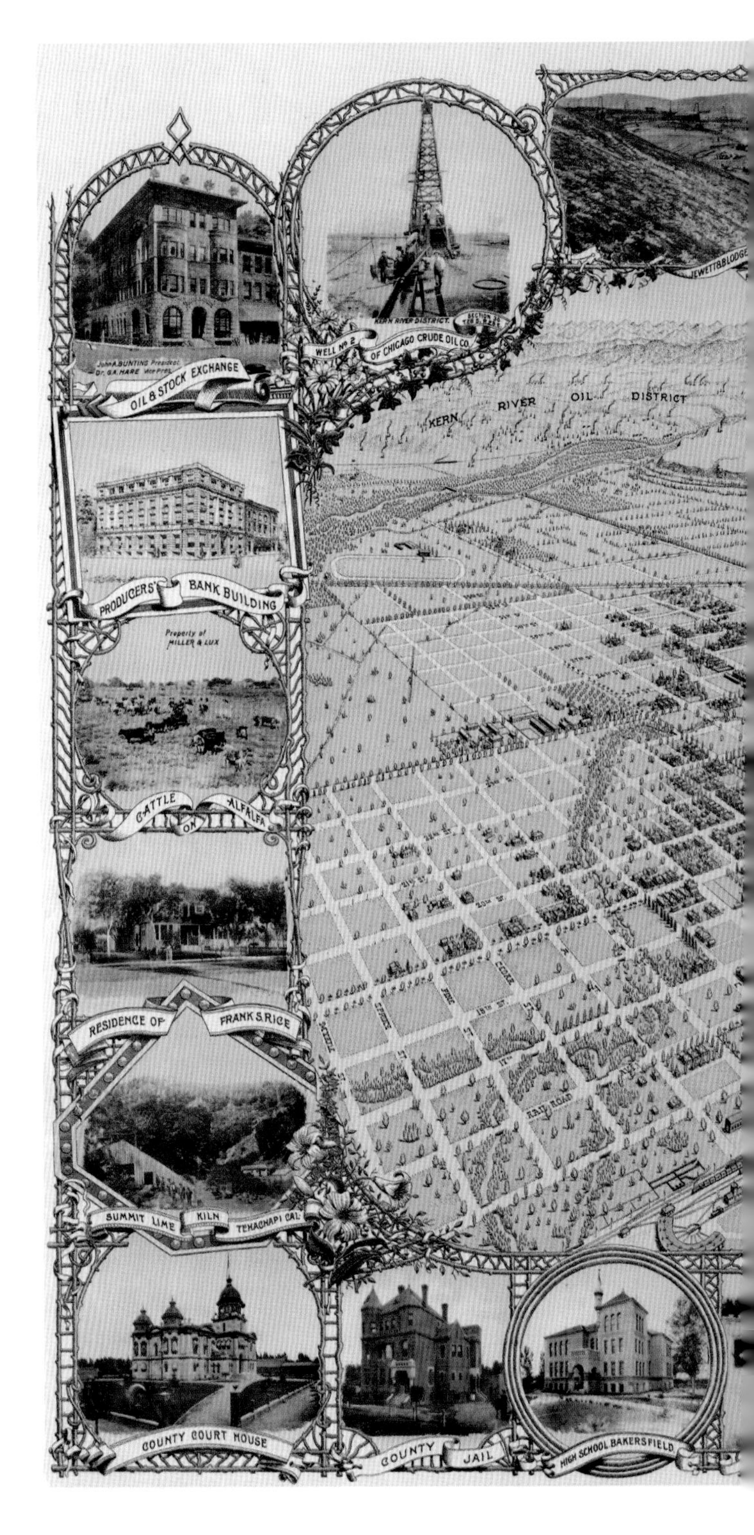

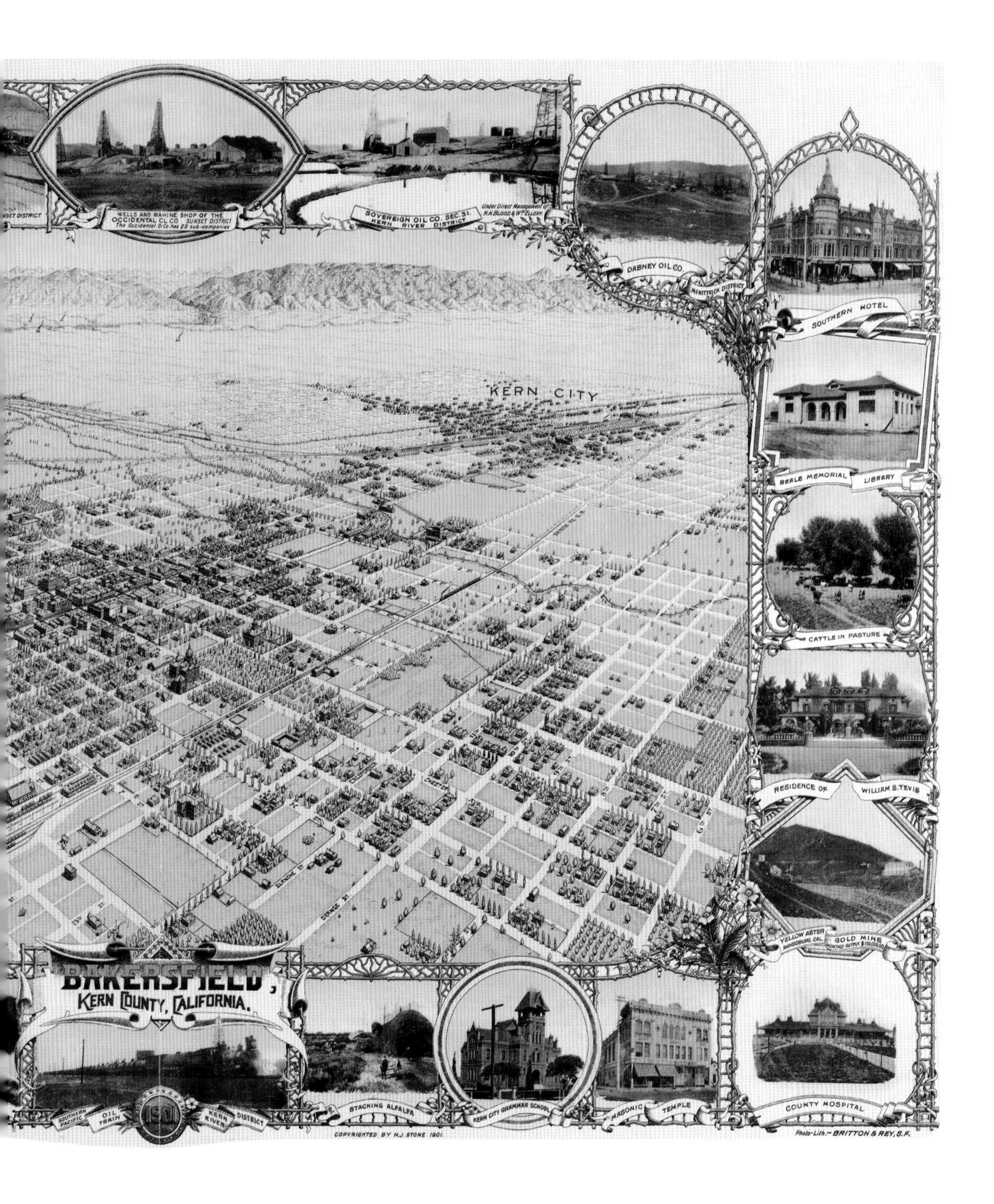

SET DISTRICT
WELLS AND MAHINE SHOP OF THE OCCIDENTAL CL CO SUNSET DISTRICT
The Occidental Oi Co. has 22 sub-companies
SOVEREIGN OIL CO. SEC. 31. KERN RIVER DISTRICT
Under Direct Management of H.H. BLOOD & Wm. ELLERY.
DABNEY OIL CO.
McKITTRICK DISTRICT
SOUTHERN HOTEL
KERN CITY
BEALE MEMORIAL LIBRARY
CATTLE IN PASTURE
RESIDENCE OF WILLIAM S. TEVIS
YELLOW ASTER GOLD MINE
BAKERSFIELD,
KERN COUNTY, CALIFORNIA.
SOUTHERN PACIFIC OIL TRAIN
1901
KERN RIVER DISTRICT
STACKING ALFALFA
KERN CITY GRAMMAR SCHOOL
MASONIC TEMPLE
COUNTY HOSPITAL
COPYRIGHTED BY N.J. STONE 1901.
Photo-Lith.~ BRITTON & REY, S.F.

S.F. HOLLANDER 405 2ND ST.
C.H. WRIGHT JEWELER 209 F ST.
REAL ESTATE & INSURANCE OFFICE OF G.R. GEORGESON
COUNTY COURT HOUSE
BUHNE'S BIG STORE
EUREKA FOUNDRY CO.
N.H. PINE President Chas. H. ELSNER Secretary
McNAMARA'S
2ND ST. ENTRANCE
W.A. McNAMARA
G.A. WALDNER Prop.
COUNTY HOSPITAL
H.R. BRUHNS
WOODLEY ISLAND
⊕ SITE OF CARNEGIE LIBRARY. COST $30,000
EUREKA
HUMBOLDT COUNTY
CALIFORNIA
THE H.D. BENDIXEN SHIPBUILDING CO.
A. COTTRELL COR. 5TH & H STS.
J.W. HENDERSON PRESIDENT N. McMILLAN SECRETARY
HUMBOLDT BAY WOOLEN MILLS
G.R. GEORGESON RESIDENCE COR. 6TH & J STS.
REDWOOD LAND & INVESTMENT CO.
PUBLISHED BY 1902 A.C. NOE & G.R. GEORGESON
COPYRIGHTED BY A.C. NOE & G.R. GEORGESON 1902
PHOTOS BY MILLER, EUREKA.

Eureka, Humboldt County, California (1902).

Upon making one of his many scientific discoveries, the ancient Greek genius Archimedes is said to have cried out, "Eureka" or "I have found it." During the gold rush, prospectors found gold in the mountains east of Humboldt Bay, and this drove the development of Eureka, which had an attractive harbor. By the time this bird's-eye view of Eureka was published in 1902, most of the gold had long since vanished. For their livelihood locals turned to lumbering the towering sequoia forests that covered these coasts in a dark green mantle. In the end the red wood of the sequoias produced far greater riches than had the gold. Much of the lumber used to build San Francisco came from this area.

Berkeley (1909).

Although not drawn to scale, this 1909 aerial view depicts Berkeley somewhat as it may have appeared when seen from the Berkeley Hills just east of the city. Located beside San Francisco Bay and adjacent to Oakland, Berkeley was once known as Ocean View. Renamed during the 1860s for Irish philosopher and poet George

Berkeley, the city was chosen as the site of the University of California, founded in 1868. The university grounds can be seen in the center foreground on the map. Berkeley and its university have hosted many world-renowned thinkers and scientists. Among these was Robert J. Oppenheimer, a key figure in the development of the first atomic bomb.

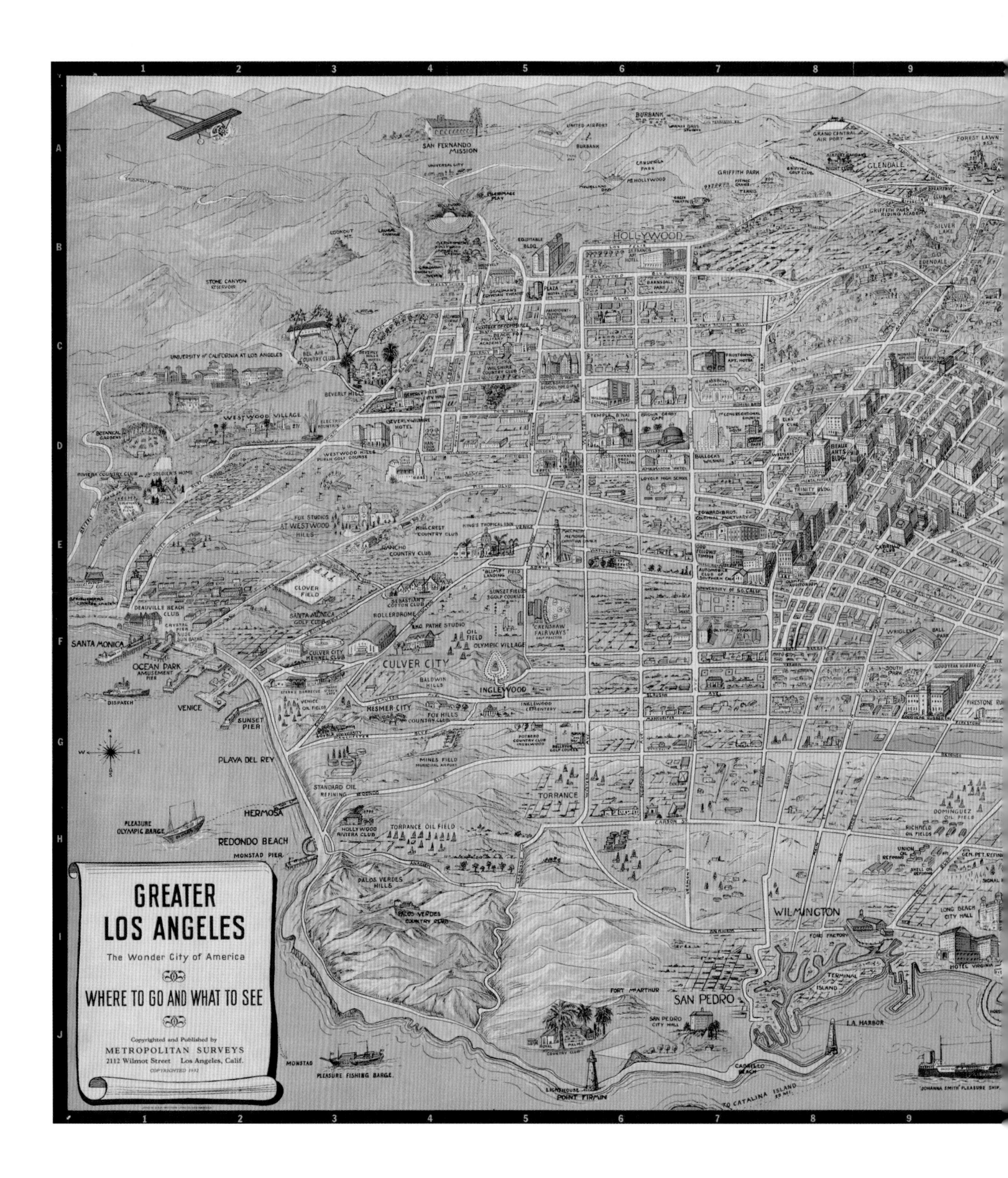
GREATER LOS ANGELES
The Wonder City of America
WHERE TO GO AND WHAT TO SEE
Copyrighted and Published by
METROPOLITAN SURVEYS
2112 Wilmot Street Los Angeles, Calif.
SAN FERNANDO MISSION
BURBANK
UNITED AIRPORT
GRAND CENTRAL AIR PORT
GLENDALE
FOREST LAWN
GRIFFITH PARK
MT. HOLLYWOOD
HOLLYWOOD
SILVER LAKE
STONE CANYON RESERVOIR
UNIVERSITY OF CALIFORNIA AT LOS ANGELES
BEL AIR COUNTRY CLUB
BEVERLY HILLS
WESTWOOD VILLAGE
BOTANICAL GARDENS
RIVIERA COUNTRY CLUB
SOLDIER'S HOME
FOX STUDIOS AT WESTWOOD HILLS
HILLCREST COUNTRY CLUB
RANCHO COUNTRY CLUB
CLOVER FIELD
DEAUVILLE BEACH CLUB
SANTA MONICA GOLF CLUB
ROLLERDROME
OLYMPIC VILLAGE
SANTA MONICA
OCEAN PARK
VENICE
CULVER CITY
BALDWIN HILLS
INGLEWOOD
MESMER CITY
FOX HILLS COUNTRY CLUB
SUNSET PIER
PLAYA DEL REY
MINES FIELD
STANDARD OIL REFINING
HERMOSA
PLEASURE OLYMPIC BARGE
REDONDO BEACH
MONSTAD PIER
HOLLYWOOD RIVIERA CLUB
TORRANCE OIL FIELD
TORRANCE
PALOS VERDES HILLS
PALOS VERDES COUNTRY CLUB
UNIVERSITY OF SO. CALIF.
WRIGLEY
BEAUX ARTS BLDG.
TRINITY BLDG.
DOMINGUEZ OIL FIELD
RICHFIELD OIL FIELDS
WILMINGTON
LONG BEACH CITY HALL
HOTEL VIRGINIA
TERMINAL ISLAND
FORT MACARTHUR
SAN PEDRO
SAN PEDRO CITY HALL
L.A. HARBOR
CABRILLO BEACH
POINT FIRMIN
TO CATALINA ISLAND
PLEASURE FISHING BARGE
'JOHANNA SMITH' PLEASURE SHIP

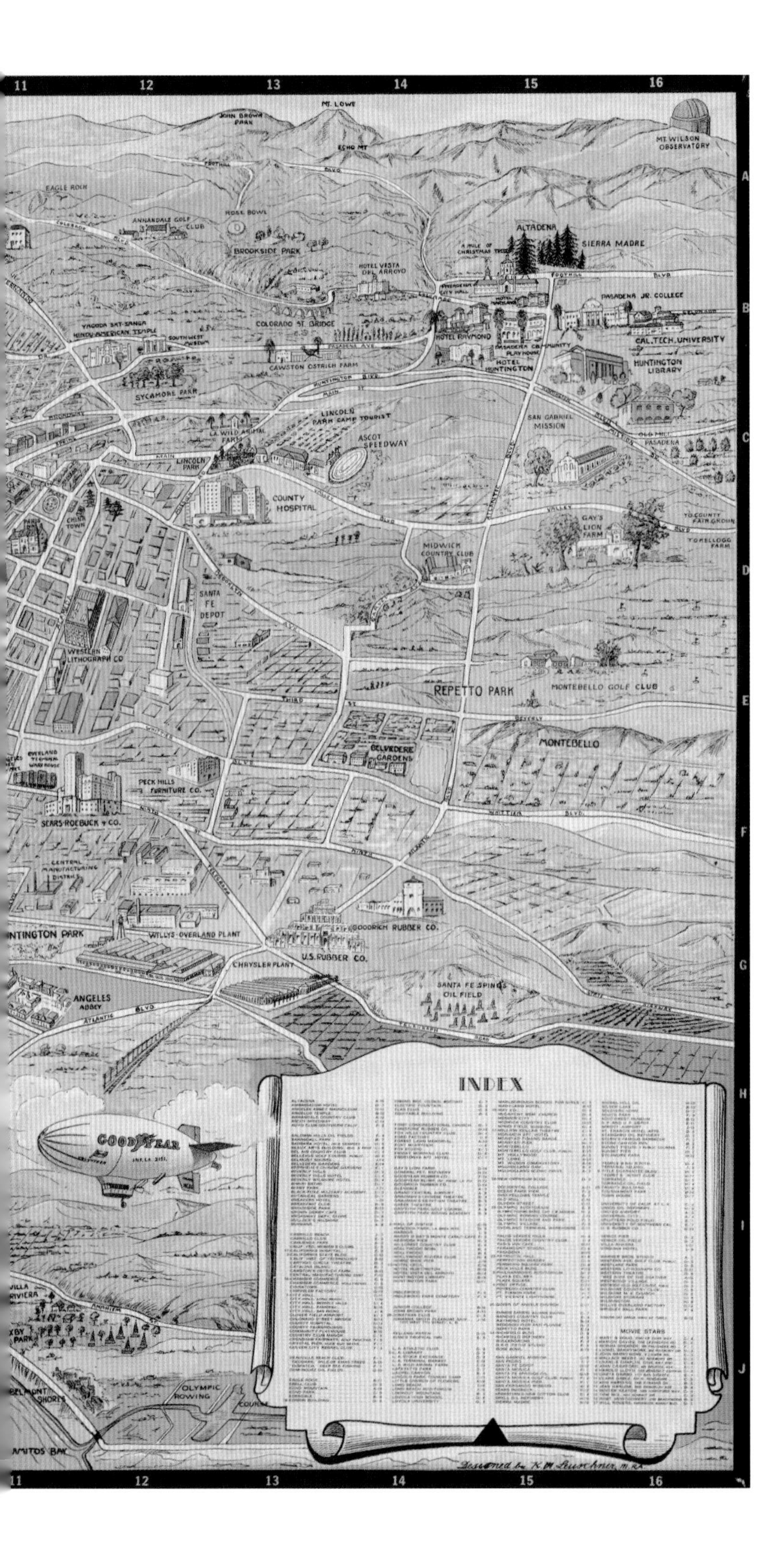

Greater Los Angeles: the wonder city of America, designed by K. M. Leuschner (1932).

A colorful projection of Greater Los Angeles details many area attractions with miniature illustrations. By the time this 1932 view was produced, Los Angeles was the largest city in California and one of the largest in the United States. Although much of America was caught in the grip of the Great Depression, Los Angeles prospered and grew tremendously during the 1930s. Many of its most famous homes and buildings were built during this period and display the distinctive art deco styling prevalent during that era.

View of Sacramento City as it appeared during the great inundation in January 1850, drawn from nature by Geo. W. Casilear & Henry Bainbridge.

Sacramento was founded in 1849 by John Sutter Jr., son of the more-famous John Sutter, who built and maintained a fort near the site of the future capital. However, the miners, merchants, and others who crowded into the newly chartered town may have been unaware that it was located in the Sacramento River floodplain. In January of 1850 the river overflowed its banks, inundating the town and flooding out almost the entire population. This view provides a sense of the severity of the flooding. Loss of life was relatively slight, given the extent of the devastation, but not so the cholera epidemic that followed in the wake of the flood. Hundreds are said to have perished. Levies were built later in an attempt to tame the rampaging river, but flooding has remained a concern for Sacramento to this day.

Legendary Disasters

California is a state of superlatives. Everyone knows that California is our nation's most populous and wealthiest state. In fact, the California economy outperforms all but a few of the world's entire nations. Blackberries are bigger and tastier in California. Peaches are juicier, lettuce is fresher, and artichokes are more succulent. Politics is also weirder in California than in most other states, and traffic jams are certainly more exasperating. Another thing the state produces on a major scale is disaster. Many of California's natural and man-made disasters are legendary.

Almost certainly the most notorious of California's many epic disasters was the great earthquake and fire that all but destroyed San Francisco in 1906. It began at 5:12 on the morning of April 18, 1906, with a rumble not unlike that of a train racing along a track. However, the roar that turned San Franciscans out of bed that Wednesday morning was not produced by man-made locomotives or railroad cars. It was the work of nature. Out on the floor of the Pacific just a few miles west of the city, a crack had opened in the earth. The rupture extended along the San Andreas Fault as far south as San Juan Bautista and as far north as Point Arena, a distance of nearly 300 miles.

The vibrations generated by this cataclysmic upheaval could be felt as far away as Los Angeles, the Great Basin region of Nevada, and central Oregon, but of course, the shaking was strongest in San Francisco. Hundreds of swaying structures collapsed outright, while thousands of others were damaged beyond all hope of repair. By the time the trembling stopped, only about a minute after it had begun, thousands of San Franciscans lay dead or seriously injured in the rubble. In order to minimize public anxiety, the official death toll was put at 478, a number that most observers at the time recognized as preposterously low.

As bad as all this was, worse was yet to come. San Francisco had been built largely of redwood. Despite its considerable strength and resistance to decay, redwood is profoundly vulnerable to fire. In the wild, sequoia trees are protected by a fire-retardant cloak of shaggy bark. That, in part, is why these giant trees sometimes live for thousands of years. But once stripped of its bark and cut into planks, the fragrant reddish wood of the coastal sequoia burns like kindling. In effect, early-twentieth-century San Francisco was a mighty bonfire waiting to be lit, and the earthquake touched it with a match. Even before the shaking stopped, hundreds of small fires had been started by ruptured gas mains, overturned oil lamps, collapsed chimneys, and red-hot coals spilled out onto tinder-dry wooden floors. Soon these individual fires began to join forces, and eventually they merged into a single great conflagration that began to consume the city, block by flaming block.

San Francisco's firefighting forces had been decapitated at the outset when Fire Chief Dennis Sullivan was struck down by falling masonry during the earthquake. Efforts to combat the blaze were further hampered as broken water mains rendered fire hoses useless in many parts of the city. Having turned the great commercial buildings, hotels, and warehouses in the business district into giant torches, the fire then marched westward into the sprawling San Francisco residential neighborhoods. Hot easterly winds, some of them whipped up by the flames themselves, created a firestorm that incinerated as many as a thousand structures every hour.

For three days frantic firefighters, their ranks increased by soldiers from the San Francisco presidio and desperate citizens struggling to save their neighborhoods, fell back before the advancing inferno. Finally, at the broad north-south thoroughfare of Van Ness Avenue, they decided to make their last stand. If the fire could not be stopped here, then nothing of the city could be saved. In hopes of creating a fire break, Army Corps of Engineers demolition experts who had been pressed into service as firemen began to dynamite houses on the east side of Van Ness. In some cases the owners were given less than ten minutes to pack their belongings and get out. Meanwhile fire crews pumped frantically to hose down the roofs of houses just west of the avenue. Not long afterward the bright red wall of fire swept down the hill toward Van Ness. People cried out and prayed as it closed in on them. The flames sputtered. Then it began to rain, and the flames were quelled.

Los Angeles, too, has endured many great earthquakes. As recently as 1994 the Northridge earthquake killed more than seventy people in and around Los Angeles. The quake did more than $20 billion damage to homes, businesses, and other structures, making it one of the costliest natural disasters in U.S. history. However, the most famous Los Angeles–area disaster was not an earthquake; it was a flood unleashed upon residents of nearby Ventura County, not so much by nature as by human pride.

Throughout the city's history, a key concern for Los Angeles and its Angelinos has been finding enough fresh water to keep faucets from running dry. During the early twentieth century, the city bought up water rights in the Owens Valley, just east of the highest Sierra peaks. Well watered by the once free-flowing Owens River, the valley was a rare natural oasis in the Southern California desert. Beginning in 1913, however, the city began to divert the Owens River, pumping all but a trickle of its sparkling waters into a 235-mile-long aqueduct. The diversion proved an immediate disaster for Owens Valley farmers and ranchers, whose wells and irrigation ditches soon dried up. Although they tried to fight Los Angeles in court, the city was able to muster more than enough legal muscle to preserve the diversion project, and the pumping continued.

Of course, once the Owens River water arrived in the Los Angeles basin, it could not be used immediately; it had to be stored until it was needed. To solve this problem, the city and its chief water engineer, William Mulholland, built the St. Francis Dam along the Santa Clara River and established an enormous reservoir. Unfortunately, for thousands of people who lived in the river canyon below the reservoir, Mulholland had done a poor job of selecting a site for his dam. The rocks underlying the massive, 195-foot-tall concrete structure were shot through with cracks and fissures. No sooner was the dam completed and its floodgates closed than the dam began to leak. Over the next two years, these leaks grew larger and more alarming, but Mulholland dismissed them as normal for a concrete dam of this size. He should have paid closer attention.

Shortly before midnight on March 12, 1928, the St. Francis Dam collapsed, sending a wall of water as high as 180 feet hurtling down the canyon. North Los Angeles, which relied on the dam for its electric power, was instantly plunged into darkness, but that was a mere inconvenience. People in the Santa Clara Canyon fared far, far worse. As the great wave closed over them, homes and entire communities were swept away. Businesses were smashed and roadways obliterated. Although an accurate count has never been widely agreed upon, many believe that at least six hundred people lost their lives in this disaster. An indirect victim of the flood was William Mulholland, who never publicly accepted responsibility for the tragedy but soon retreated into a hermitlike retirement.

Earthquakes, tsunamis, mudslides, fires, floods, explosions, toxic spills, plane crashes, riots, and much more—California's disasters have certainly piled up over the years. Sometimes, after a fresh calamity is shown on the network news, Californians receive calls from their friends in the East. "Why would you want to live in such a place?" their friends ask. This tends to annoy the Californians, who may answer as follows: "These things don't happen *everywhere* in California, and anyway, the weather makes it worth the risk."

BUILDING PERMITS ISSUED SINCE FIRE TO DATE

DESCRIPTION	NO.	VALUE
CLASS A	63	$16,452,000
" B	95	7,036,671
" C	1,097	33,547,219
FRAME	8,817	37,139,694
ALTERATIONS AND REPAIRS	4,198	8,010,933
TOTAL	14,270	$102,186,517

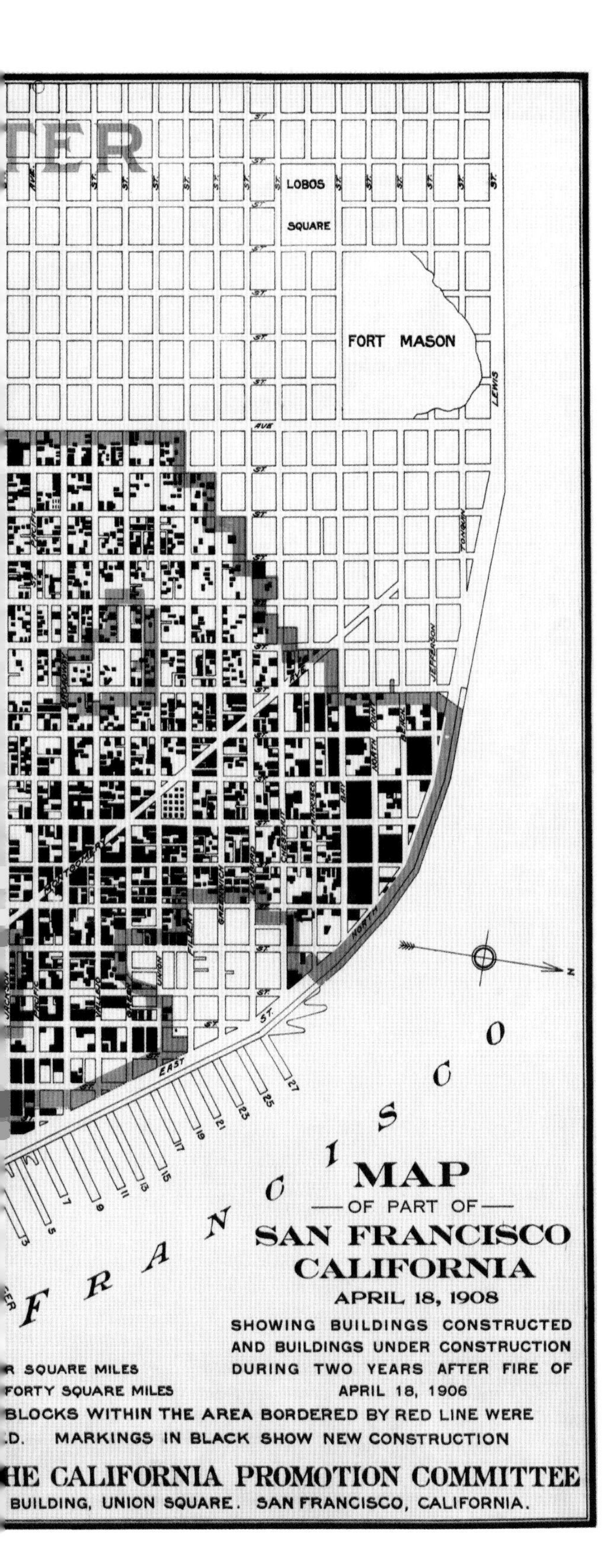

Map of part of San Francisco, California, April 18, 1908: showing buildings constructed and buildings under construction during two years after fire of April 18, 1906. Punnett Brothers. San Francisco: California Promotion Committee (1908).

The 1906 San Francisco earthquake and fire obliterated about half the city. While the earthquake knocked down some structures and severely weakened many others, it was the fire that did most of the damage. It is said that at its height the blaze consumed wooden structures at the rate of more than a thousand per hour. The disaster took an estimated three thousand lives and destroyed twenty-five thousand homes and buildings. Financial losses were put at $400 million, a fantastic sum for the early twentieth century. Yet as this map detailing two years of reconstruction effort demonstrates, the city was quick to recover.

MAGNITUDE 5.0 AND GREATER EARTHQUAKES

Earthquake No.	Date	Time (Universal Time Coordinated)	Latitude (N.)	Longitude (W.)	Depth (km)	Magnitude (Richter scale)
1	28 Nov 1974	23:01:24.56	36.9202	121.4663	6.11	5.37
2	6 Aug 1979	17:05:22.91	37.1042	121.5127	5.93	5.80
3	24 Jan 1980	19:00:08.63	37.8367	121.7708	14.72	5.23
4	24 Jan 1980	19:01:01.53	37.8116	121.7772	7.07	5.10
5	27 Jan 1980	02:33:35.32	37.7493	121.7062	14.69	5.18
6	23 Jan 1984	05:40:19.93	36.3650	121.8878	6.19	5.12
7	24 Apr 1984	21:15:18.75	37.3095	121.6787	8.74	5.82
8	26 Jan 1986	19:20:50.93	36.8040	121.2847	8.72	5.50
9	31 Mar 1986	11:55:39.80	37.4786	121.6858	9.17	5.70
10	20 Feb 1988	08:39:07.24	36.7957	121.3117	9.82	5.03
11	13 June 1988	01:45:38.52	37.3923	121.7415	9.71	5.30
12	27 June 1988	18:43:22.32	37.1288	121.8962	13.18	5.30
13	8 Aug 1989	08:13:27.36	37.1468	121.9282	14.23	5.40
14	18 Oct 1989	00:04:15.20	37.0362	121.8802	17.29	7.00
15	18 Oct 1989	00:41:23.70	37.1873	122.0548	16.78	5.10
16	18 Apr 1990	13:53:51.30	36.9320	121.6577	5.68	5.75
17	18 Apr 1990	15:46:03.44	36.9583	121.6848	6.67	5.36
18	18 Jan 1993	06:29:35.00	37.0182	121.4630	7.90	5.10
19	11 Aug 1993	22:33:04.00	37.3120	121.6793	9.35	5.00
20	23 Apr 1995	08:41:36.83	36.6025	121.2015	7.84	5.14
21	12 Aug 1998	14:10:25.12	36.7533	121.4622	9.13	5.35
22	3 Sept 2000	08:36:30.09	38.3798	122.4127	10.12	5.17

Earthquakes and faults in the San Francisco Bay Area (1970–2003), by Benjamin M. Sleeter . . . [et al.] (2004).
There is absolutely no reason to believe that San Francisco won't someday be visited once again with a natural disaster as deadly and destructive as the one that struck on April 18, 1906. In fact, as this recent U.S. Geological Survey map strongly suggests, it is sure to happen. The entire San Francisco area is shot through with subterranean faults that will sooner or later give way and subject the city to another 1906-style shaking. Among these cracks is the notorious San Andreas Fault, which runs directly beneath the streets of the city.

CSSC No. 03-02

STATE OF CALIFORNIA - GRAY DAVIS, GOVERNOR

Earthquake Shaking Potential for California

Spring, 2003

This map shows the relative intensity of ground shaking and damage in California from anticipated future earthquakes. Although the greatest hazard is in the areas of highest intensity as shown on the map, no region within the state is immune from potential for earthquake damage. Expected damages in California in the next 10 years exceed $30 billion.

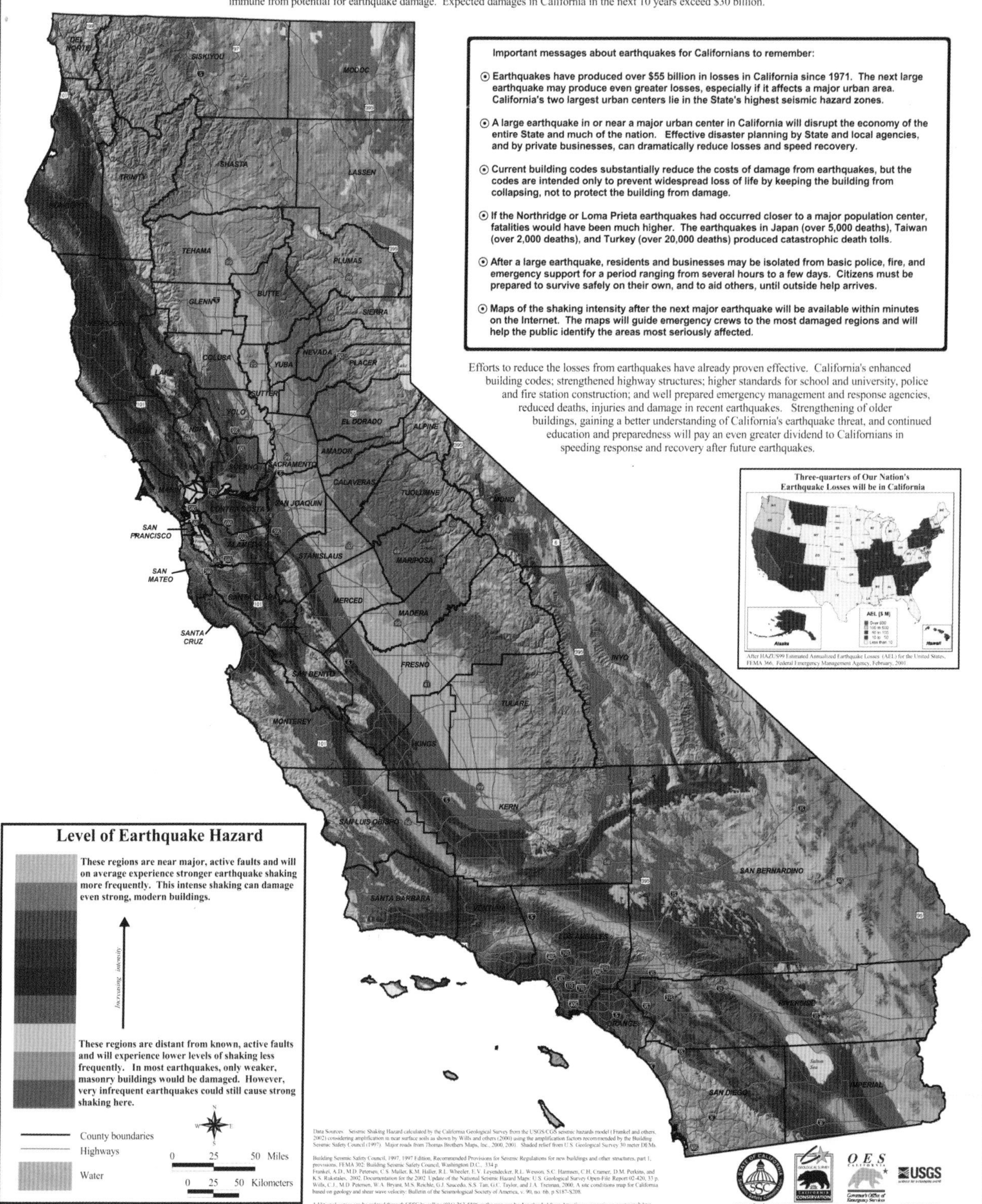

Important messages about earthquakes for Californians to remember:
Earthquakes have produced over $55 billion in losses in California since 1971. The next large earthquake may produce even greater losses, especially if it affects a major urban area. California's two largest urban centers lie in the State's highest seismic hazard zones.
A large earthquake in or near a major urban center in California will disrupt the economy of the entire State and much of the nation. Effective disaster planning by State and local agencies, and by private businesses, can dramatically reduce losses and speed recovery.
Current building codes substantially reduce the costs of damage from earthquakes, but the codes are intended only to prevent widespread loss of life by keeping the building from collapsing, not to protect the building from damage.
If the Northridge or Loma Prieta earthquakes had occurred closer to a major population center, fatalities would have been much higher. The earthquakes in Japan (over 5,000 deaths), Taiwan (over 2,000 deaths), and Turkey (over 20,000 deaths) produced catastrophic death tolls.
After a large earthquake, residents and businesses may be isolated from basic police, fire, and emergency support for a period ranging from several hours to a few days. Citizens must be prepared to survive safely on their own, and to aid others, until outside help arrives.
Maps of the shaking intensity after the next major earthquake will be available within minutes on the Internet. The maps will guide emergency crews to the most damaged regions and will help the public identify the areas most seriously affected.
Efforts to reduce the losses from earthquakes have already proven effective. California's enhanced building codes; strengthened highway structures; higher standards for school and university, police and fire station construction; and well prepared emergency management and response agencies, reduced deaths, injuries and damage in recent earthquakes. Strengthening of older buildings, gaining a better understanding of California's earthquake threat, and continued education and preparedness will pay an even greater dividend to Californians in speeding response and recovery after future earthquakes.
Three-quarters of Our Nation's Earthquake Losses will be in California
AEL ($ M)
Alaska
Hawaii
After HAZUS99 Estimated Annualized Earthquake Losses (AEL) for the United States, FEMA 366, Federal Emergency Management Agency, February, 2001.
DEL NORTE
SISKIYOU
MODOC
SHASTA
TRINITY
LASSEN
TEHAMA
PLUMAS
GLENN
BUTTE
SIERRA
COLUSA
YUBA
NEVADA
PLACER
LAKE
SUTTER
YOLO
EL DORADO
ALPINE
AMADOR
SACRAMENTO
CALAVERAS
TUOLUMNE
MONO
SAN JOAQUIN
CONTRA COSTA
SAN FRANCISCO
SAN MATEO
STANISLAUS
MARIPOSA
MERCED
MADERA
SANTA CRUZ
SAN BENITO
FRESNO
INYO
TULARE
MONTEREY
KINGS
KERN
SAN LUIS OBISPO
SAN BERNARDINO
SANTA BARBARA
RIVERSIDE
ORANGE
SAN DIEGO
IMPERIAL
Level of Earthquake Hazard
These regions are near major, active faults and will on average experience stronger earthquake shaking more frequently. This intense shaking can damage even strong, modern buildings.
Increasing intensity
These regions are distant from known, active faults and will experience lower levels of shaking less frequently. In most earthquakes, only weaker, masonry buildings would be damaged. However, very infrequent earthquakes could still cause strong shaking here.
County boundaries
Highways
Water
0 25 50 Miles
0 25 50 Kilometers
Data Sources: Seismic Shaking Hazard calculated by the California Geological Survey from the USGS/CGS seismic hazards model (Frankel and others, 2002) considering amplification in near surface soils as shown by Wills and others (2000) using the amplification factors recommended by the Building Seismic Safety Council (1997). Major roads from Thomas Brothers Maps, Inc., 2000, 2001. Shaded relief from U.S. Geological Survey 30 meter DEMs.
Building Seismic Safety Council, 1997, 1997 Edition, Recommended Provisions for Seismic Regulations for new buildings and other structures, part 1, provisions, FEMA 302: Building Seismic Safety Council, Washington D.C., 334 p.
Frankel, A.D., M.D. Petersen, C.S. Muller, K.M. Haller, R.L. Wheeler, E.V. Leyendecker, R.L. Wesson, S.C. Harmsen, C.H. Cramer, D.M. Perkins, and K.S. Rukstales, 2002, Documentation for the 2002 Update of the National Seismic Hazard Maps: U.S. Geological Survey Open-File Report 02-420, 33 p.
Wills, C.J., M.D. Petersen, W.A. Bryant, M.S. Reichle, G.J. Saucedo, S.S. Tan, G.C. Taylor, and J.A. Treiman, 2000, A site conditions map for California based on geology and shear wave velocity: Bulletin of the Seismological Society of America, v. 90, no. 6b, p S187-S208.
Additional copies can be ordered through CSSC by calling (916) 263-5506 or the map can be downloaded from http://www.seismic.ca.gov/sscpub.htm.
OES
USGS
www.seismic.ca.gov
www.consrv.ca.gov
www.oes.ca.gov
www.usgs.gov

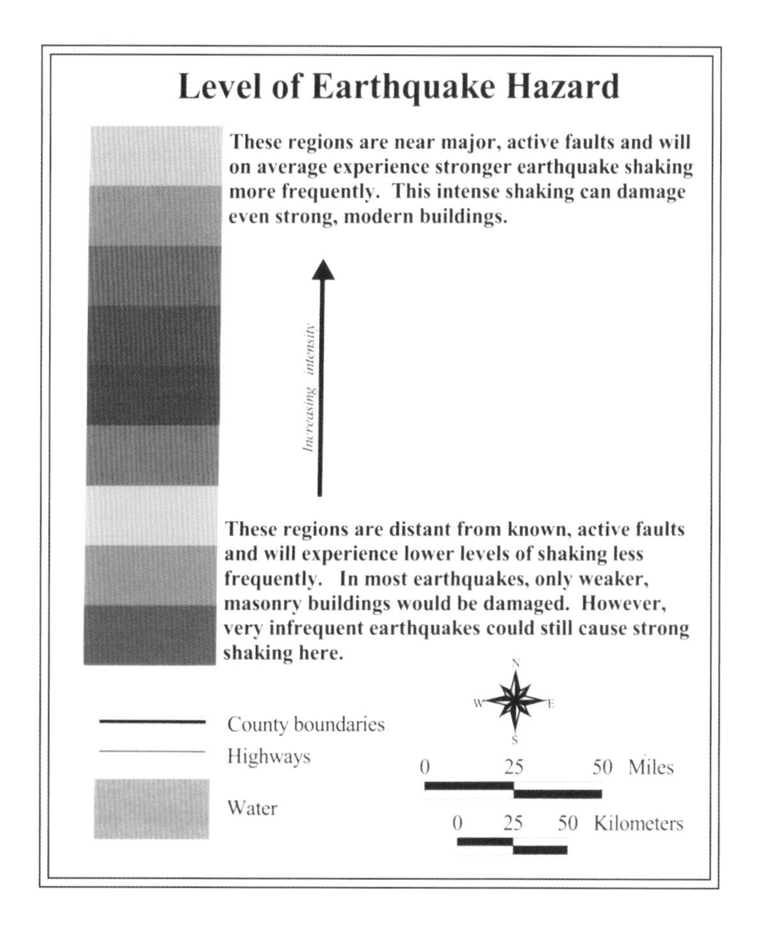

Level of Earthquake Hazard
These regions are near major, active faults and will on average experience stronger earthquake shaking more frequently. This intense shaking can damage even strong, modern buildings.
Increasing intensity
These regions are distant from known, active faults and will experience lower levels of shaking less frequently. In most earthquakes, only weaker, masonry buildings would be damaged. However, very infrequent earthquakes could still cause strong shaking here.
N
W
E
S
County boundaries
Highways
Water
0 25 50 Miles
0 25 50 Kilometers

Earthquake shaking potential for California, Spring, 2003, California Seismic Safety Commission (2002).

Earthquakes can strike at anytime almost anywhere in California. Scientists have tried for decades to develop techniques for predicting quakes, especially massive and highly destructive ones, such as those that struck the Bay Area in 1906 and 1989. Unfortunately, they've had limited success. The Seismic Safety Commission map shown here indicates areas of unusually high seismic activity. Not surprisingly, the highest activity, marked in light purple, traces the course of the San Andreas Fault, which runs some 800 miles from the Salton Sea in Southern California to Cape Mendocino north of San Francisco.

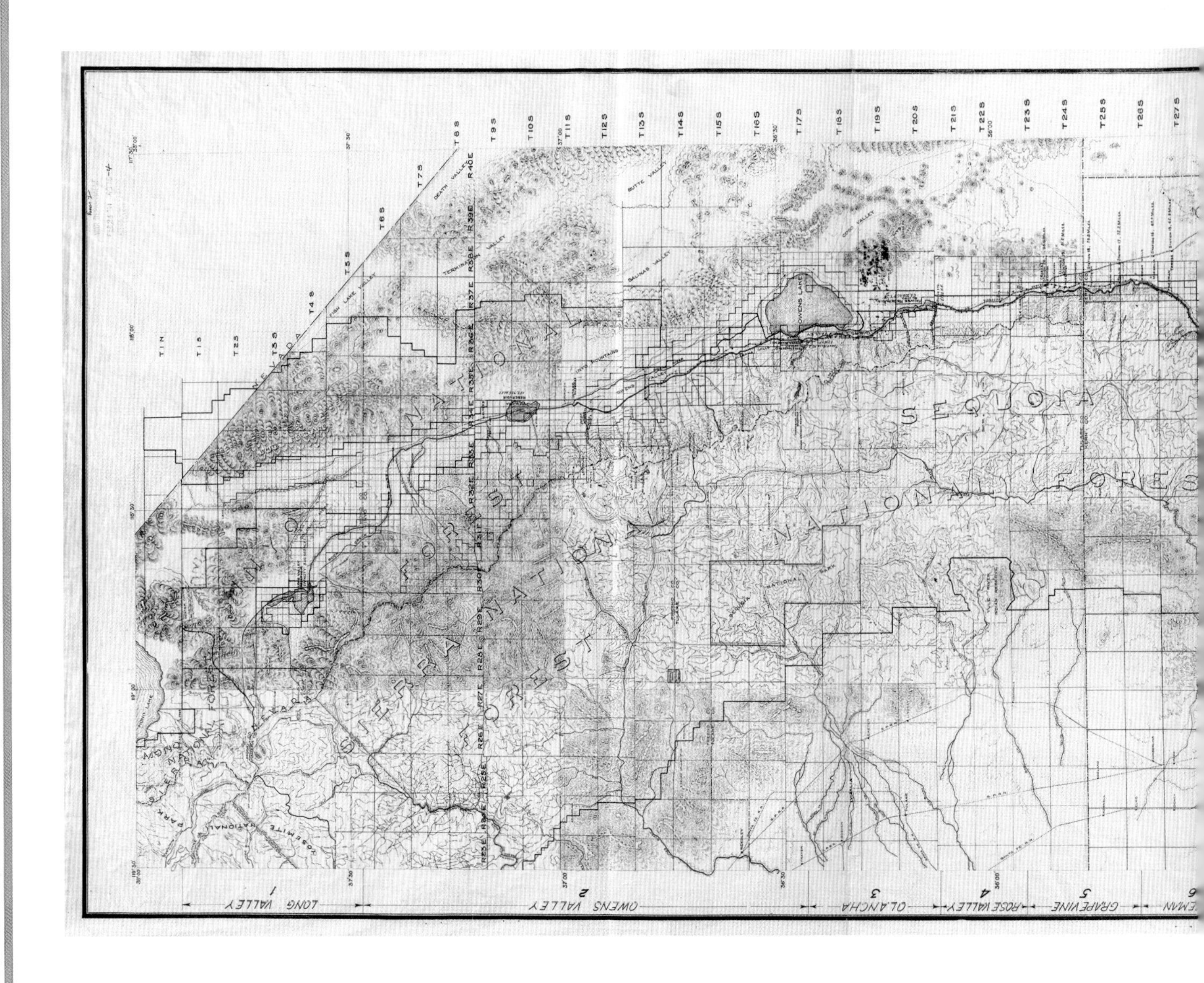

Topographic map of the Los Angeles aqueduct and adjacent territory,
Water Department of the City of Los Angeles, Cal. Board of Commissioners (1908).

The Los Angeles Aqueduct shown here was one of the most ambitious—and in some ways destructive—water-diversion projects in the history of the West. During the early twentieth century, the City of Los Angeles began quietly acquiring water rights in the Owens Valley and in 1913 started pumping Owens River water through

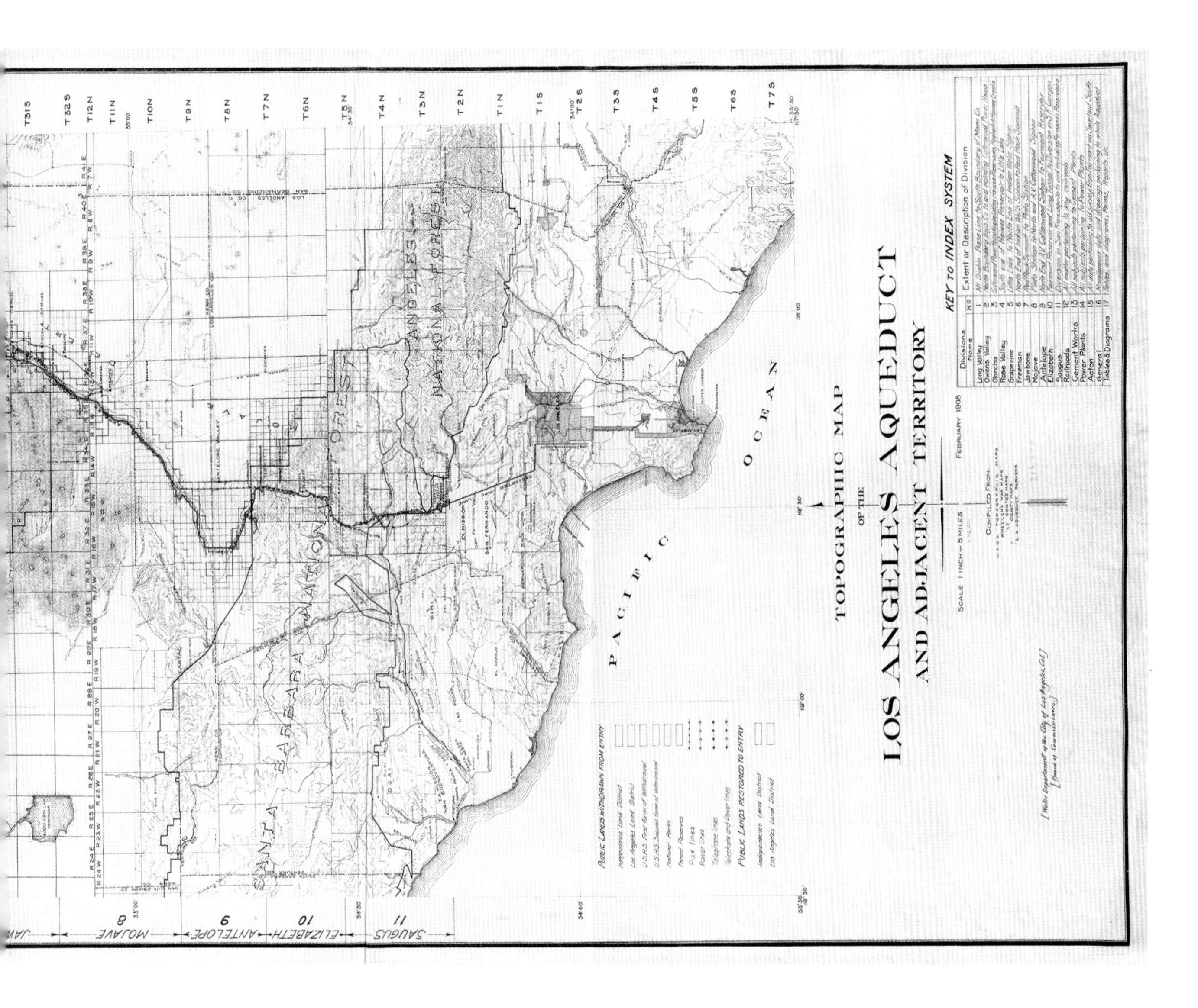

an enormous system of pipes and canals to reservoirs near Los Angeles more than 200 miles to the west. The diversion devastated the Owens Valley, where the once-verdant fields of local farms and ranches were turned to desert. The diversion also led to disaster in the Los Angeles area when the St. Francis Dam collapsed, sending a 180-foot-high wall of aqueduct water surging through the Santa Clara River Valley. Thousands of homes and other buildings were destroyed and more than six hundred lives lost in the calamity.

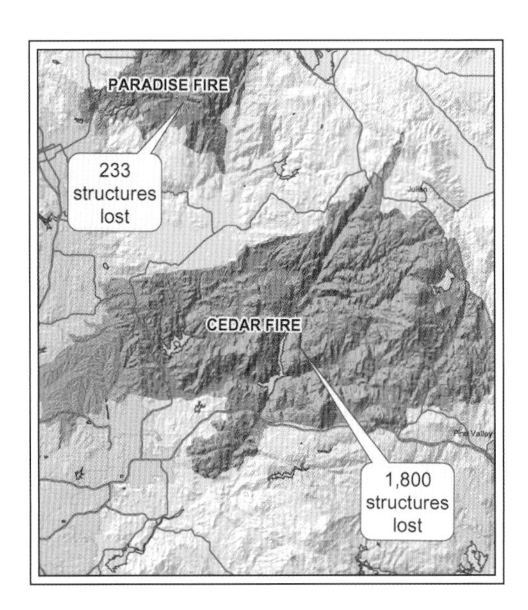

INCIDENT NAME	SIZE(ACRES)	% CONTAINME
CEDAR	251,000	15
GRAND PRIX	91,207	40
PADUA	10,466	90
PARADISE	49,800	20
PIRU	68,022	30
OLD	47,960	10
MOUNTAIN	10,331	100
ROBLAR 2	8,592	85
SIMI	105,665	40

Source: NFIC

Southern California Wildfire Overview: 10/30/03, MAST, Mountain Area Safety Taskforce.
Another type of disaster that often torments Californians is wildfire. Like earthquakes, wildfires are naturally occurring. Often they are started by lightning, and since much of the state is tinder dry in summer, especially during drought years, there is little to prevent a wildfire from marching across thousands of acres. All too often homes and entire communities are swept away by the flames. This California Department of Forestry map is marked in red to show parts of Southern California burned over in 2003, one of the state's worst fire years in recent memory. The fire at the lower right (close-up above) was in San Diego County; in less than a week it killed sixteen people and destroyed 2,437 homes.

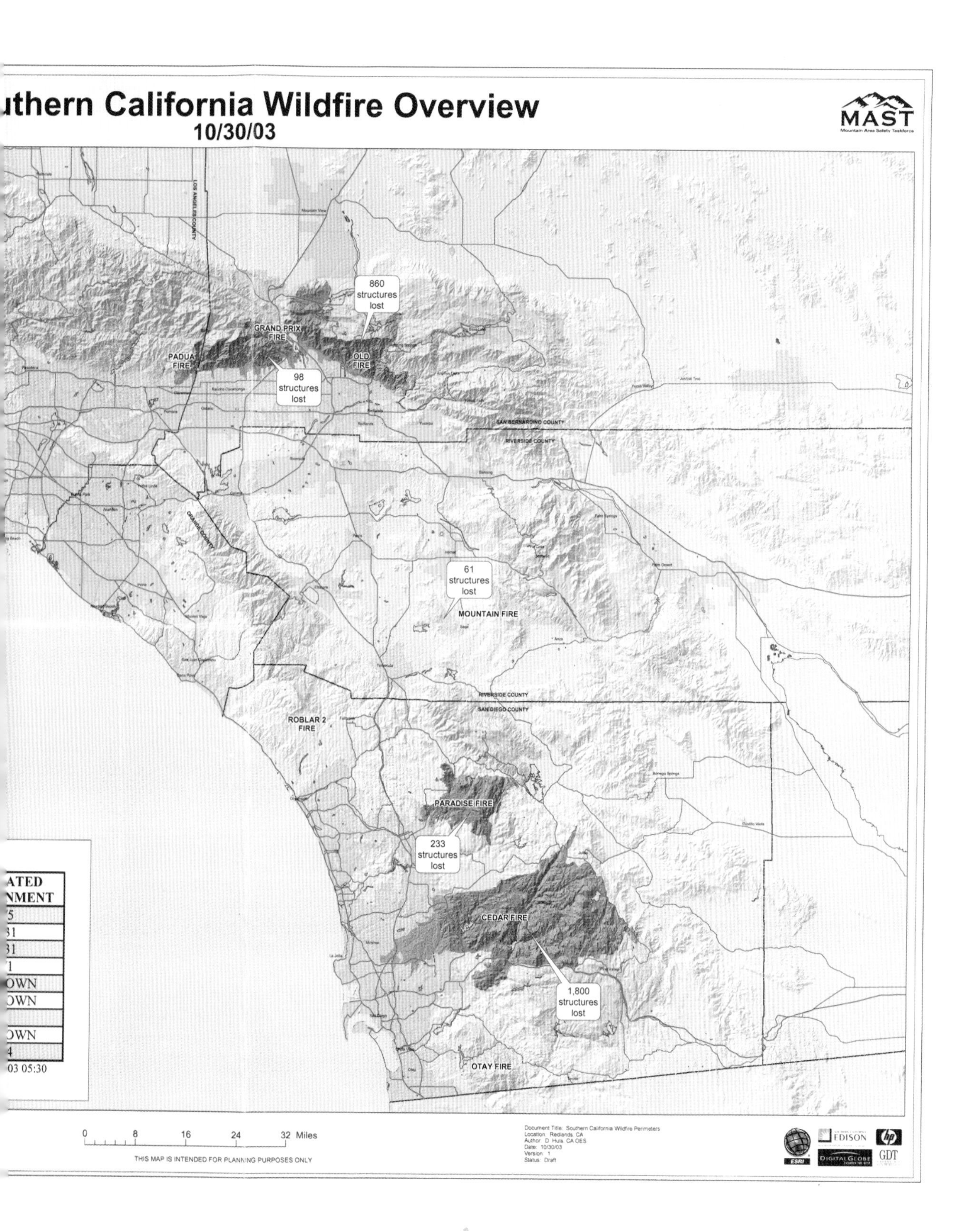

uthern California Wildfire Overview
10/30/03
MAST
Mountain Area Safety Taskforce
860 structures lost
GRAND PRIX FIRE
PADUA FIRE
OLD FIRE
98 structures lost
SAN BERNARDINO COUNTY
RIVERSIDE COUNTY
LOS ANGELES COUNTY
ORANGE COUNTY
61 structures lost
MOUNTAIN FIRE
RIVERSIDE COUNTY
SAN DIEGO COUNTY
ROBLAR 2 FIRE
PARADISE FIRE
233 structures lost
CEDAR FIRE
1,800 structures lost
OTAY FIRE
ATED
NMENT
5
31
31
1
OWN
OWN
OWN
4
03 05:30
0 8 16 24 32 Miles
THIS MAP IS INTENDED FOR PLANNING PURPOSES ONLY
Document Title: Southern California Wildfire Perimeters
Location: Redlands, CA
Author: D. Huls, CA OES
Date: 10/30/03
Version: 1
Status: Draft
ESRI
EDISON
DIGITALGLOBE
GDT

THE SIERRA
CAPE HORN
SOIL
TEMPERATURE
MEAN ANNUAL
BIG TREES
MODOC
DEL NORTE
SISKIYOU
TRINITY
SHASTA
LASSEN
HUMBOLDT
TEHAMA
PLUMAS
MENDOCINO
BUTTE
COLUSA
SIERRA
YUBA
NEVADA
LAKE
SUTTER
PLACER
YOLO
ELDORADO
SONOMA
NAPA
ALPINE
MARIN
SOLANO
CALAVERAS
MONO
SAN JOAQUIN
CONTRA COSTA
TUOLUMNE
ALAMEDA
STANISLAUS
MARIPOSA
SANTA CLARA
MERCED
INYO
SAN BENITO
FRESNO
TULARE
MONTEREY
SAN LUIS OBISPO
KERN
SAN-BERNARDINO
SANTA BARBARA
VENTURA
LOS ANGELES
SAN DIEGO
NEW YORK
NEW JERSEY
MASSACHUSETTS
VERMONT
MAINE
NEW HAMPSHIRE
OHIO
AREA
Area of the ten states quoted 153.130 Sq. M
Area of California - 153.600 -
S. McD. Johnstone
DICKMAN-JONES CO. LITH. S.F.
THE UNIQUE MAP OF CALIFORNIA
COPYRIGHTED.
ANNUAL RAINFALL IN INCHES
Published under the auspices of the Southern Pacific Company, and the State Board of Trade of California
EUREKA

Cultural Landscape

To people from elsewhere California culture is a mystery. That should come as no surprise, since Californians have a very difficult time understanding their own culture, and why wouldn't they? Influenced by the languages and customs of Native Americans, Latin Americans, Asians, Africans, and Polynesians, as well as Northern and Southern Europeans, California is one of the most culturally diverse places on the planet. Accentuating this diversity is the fact that Californians tend to revel in their own cultural heritage rather than assimilate the ways of Middle America. For these reasons it is much easier to describe California's culture in terms of what the state has given to the rest of the nation and the world rather than what they have given to California.

California seems to produce a great writer every few years. Among those who have penned classics, the name Jack London stands out. No well-read person could claim to have never turned the pages of London's *Call of the Wild*, published in 1903. This gem of a young person's novel recounts the adventures of a dog named Buck in the frozen Yukon. It is probably worth noting that Buck was not from Canada or Alaska but started out his life in California.

Author Gertrude Stein started out in California as well but spent much of her life outside the state. The explanation for this may lie in Stein's famous description of Oakland, her hometown. "There is no there there," she said. Stein lived much of her productive and

The unique map of California (1888).

Over the years the California lifestyle has been a golden dream for many. People come here not just because they hope to find gold, silver, or oil but because they are seeking more attractive surroundings and a better day-to-day existence. California's weather, scenery, and unique natural wonders are inducement enough for most. This artfully illustrated map celebrates many of the state's most attractive features, among them palm trees, sequoias, waterfalls, Mount Shasta, open roads, fine architecture, fascinating cities, and much more. An inset at the center of the page depicts Yosemite Valley.

highly creative life in Paris, where she entertained other literary giants, the likes of Scott Fitzgerald and Ernest Hemingway. She may be best known for her poetic *Autobiography of Alice B. Toklas,* which was published in 1933.

John Steinbeck grew up and began his writing career in Salinas, a quiet corner of California's agricultural heartland. Steinbeck listened intently to stories told by the migrant workers who labored in the vast fields stretching out in all directions from Salinas. He became convinced that from these often poignant tales could be distilled the essence of the American spirit. One such story he turned into a novel about a pair of Depression-era field hands, George and his friend, a retarded giant named Lennie, whose dreams of a better life went tragically astray. As a matter of fact, the novel itself went astray—the Steinbeck family dog ate the original handwritten manuscript. Steinbeck rewrote it from memory and submitted to his publisher. When *Of Mice and Men* was finally released in 1937, it quickly sold more than a hundred thousand copies; so many, in fact, that the book's success frightened its publicity-shy author. Many of those who have been exposed to this book—countless millions have been fortunate enough to read it—consider it the Great American Novel. Interestingly, one reason sometimes offered for denying *Of Mice and Men* this accolade is that it may rightfully belong instead to *The Grapes of Wrath*—another Steinbeck novel.

As George and Lennie could have testified, the most cherished dreams of life are seldom fully realized—except of course in Hollywood. For more than a century, the California movie industry has been repackaging the dreams of Americans, indeed of everyone in the whole world. When those dreams reappear on the silver screen, they are better structured and more believable. They are also a bit more polished and pretty. After all, it's hard to imagine oneself being as handsome as Clark Gable or as sexy as Marilyn Monroe, except perhaps for an hour or two while sitting in a theater or watching an old movie on cable.

Early American movies had been made at studios in the East, but the weather was often cloudy there, which made outdoor filming a hit-or-miss affair. The beginnings of the California movie industry reach back to 1908, when producer William Selig arrived in Los Angeles to film the outdoor scenes of *The Count of Monte Cristo.* He liked the bright sunny weather he encountered in Southern California and stayed on to film all of *In the Sultan's Power,* the first movie to be shot in its entirety in Los Angeles. Other producers and directors soon joined Selig. The industry put down roots, and it has flourished ever since in and around Hollywood, a district within the city limits of Los Angeles.

Movies may very well remind people of their own best—or worst—qualities. However, there are those who argue that the surest way to gain personal insight is through exposure to nature and the outdoors. Among the first to put forward this idea was naturalist John Muir, whose voluminous writings and tireless efforts would help preserve the beauty of the California Sierras in general and Yosemite National Park in particular.

Muir was not quite thirty when he arrived in San Francisco in 1868 and set out at once for the region that is now designated as Yosemite National Park. Having read about the wonders of this place, Muir yearned to see it, but it took him almost two weeks to reach it using his customary means of transport—his own two feet. When he arrived, Muir was stunned by the towering Yosemite rocks and waterfalls. From then on he made his home in Northern California, either living near Yosemite or in San Francisco only a hundred miles or so to the west.

Muir was instrumental in the establishment of Yosemite as a national park in 1890. Two years later he helped found the Sierra Club, one of the nation's oldest and most influential conservationist organizations. In 1903 he accompanied President Theodore Roosevelt on a two-man hike into the Yosemite backcountry, an experience that helped focus the president's own conservationist ideas. Acting on the advice of Muir, Roosevelt placed the park under direct federal control, thus protecting it from possible exploitation and damage by state and local politics.

California has always been a challenge to conservationists, some of whom nowadays prefer to call themselves environmentalists. There is so much natural magnificence and beauty here to preserve—grand Sierra canyons, three-thousand-year-old coastal redwoods, giant mountain sequoias as big around as a house, desert mountain cliffs that drop directly into the Pacific, seals, sea otters, pelicans, condors, cougars—in fact, an entire continent's worth of unique natural features and wild creatures. Fortunately, there are many here who believe that all this can be preserved without giving up what California has offered people for centuries: a place where they can give clearer shape to their dreams and, if they are able, to live them.

The
John Steinbeck
Map of
America
Salinas
Fremont Peak
Santa Cruz
Watsonville
Moss Landing
Castroville
Spreckles
Chualar
Mt. Toro
Seaside
Fort Ord
Santa Lucia Mts.
Jolon
Monterey
Pacific Grove
Carmel Valley
Carmel
Point Lobos

John Steinbeck Map of America (1986). Courtesy of Molly Maguire and Aaron Silverman.
Pages from a 1986 book called *Language of the Land* provide a glimpse of America as Nobel Prize–winning author John Steinbeck saw it. Images from Steinbeck's *The Grapes of Wrath* and *East of Eden* and other classics can be found among the illustrations. Many of Steinbeck's most evocative characters are based on field hands the author met while growing up in agricultural Salinas. Steinbeck felt close to California and the many dispossessed people who called it home. For them the American dream was hardly distinguishable from survival.

Raymond Chandler Mystery Map of Los Angeles (1985). Courtesy of Molly Maguire and Aaron Silverman.

Pulp fiction writer Raymond Chandler may not have scaled the literary heights achieved by Jack London or John Steinbeck, but he entertained millions with his popular crime stories. Many of his best plots revolved around Philip Marlowe, a tough LA detective who was no gentler with women than he was with murder suspects. Many continue to read Chandler favorites such as *The Big Sleep* and *The Long Goodbye*. A page from the book *Language of the Land* shows many of the locations described in Chandler's novels. Some compare Chandler's Marlowe to Dashiell Hammett's Sam Spade, a San Francisco gumshoe.

LAKE ARROWHEAD
CRESTLINE
HIGHWAY
HOLLYWOOD
BOULEVARD
NOBLE
ENCANTO
DRIVE
OAK
KNOLL
DRIVE
WENTWORTH
CANYON
LAUREL
AVENUE
FRANKLIN
AVENUE
WILCOX
BOULEVARD
FRANKLIN
AVENUE
STREET
PLACE
GRAMERCY
AVENUE
HOLLYWOOD
BOULEVARD
VINE
SUNSET
BOULEVARD
KENMORE
WESTERN
SIERRA
BONITA
LA BREA
SANTA
MONICA
BOULEVARD
LAS PALMAS
CAHUENGA
HIGHLAND
DOHENY
BOULEVARD
DRIVE
LA CIENEGA
BOULEVARD
LOS ANGELES
BRYSON
RAMPART
BOULEVARD
ALVARADO
COURT
STREET
TEMPLE
STREET
OLIVE
BROADWAY
CENTRAL
BAUCHET
6th
WILSHIRE
BOULEVARD
STREET
7th
SPRING
ALAMEDA
8th
9th
HOOVER
TOBERMAN
PICO
BOULEVARD
AVENUE
STREET
54th
STREET
CENTRAL
AVENUE
MEXICO
BOULEVARD
STREET
PICO
OCEAN PARK
WAY

HOLLYWOOD
STARLAND
OFFICIAL MOVIEGRAPH OF THE LAND OF STARS - WHERE THEY LIVE - WHERE THEY WORK AND WHERE THEY PLAY
Copyrighted by DON BOGGS
BOX 1030
HOLLYWOOD, CALIF.
OCT 20 1937
10994
SAN FERNANDO VALLEY
BURBANK
TOLUCA LAKE
GLENDALE
PASADENA
SANTA ANITA
LOS ANGELES
HOLLYWOOD
BEVERLY HILLS
BEL-AIR
MALIBU
WESTWOOD VILLAGE
SANTA MONICA
CATALINA ISLAND
WILSHIRE BLVD.
SUNSET BLVD.
HOLLYWOOD BLVD.
ROOSEVELT HOTEL
GRAUMAN'S CHINESE THEATRE
HOLLYWOOD HOTEL
WARNER BROTHER'S FIRST NATIONAL STUDIOS
UNIVERSAL STUDIOS
WALT DISNEY STUDIOS
ROSE BOWL
HUNTINGTON LIBRARY
PLANETARIUM
GREEK THEATRE
BEVERLY HILLS HOTEL
BEVERLY WILSHIRE
COTTON CLUB
U.C.L.A.
BRENTWOOD COUNTRY CLUB
DOUGLAS AIRCRAFT CORP.
SANTA MONICA GUN CLUB
WEST SIDE TENNIS CLUB
CALIFORNIA COUNTRY CLUB
HILLCREST COUNTRY CLUB
SELZNICK INTERNATIONAL PICTURES, INC.
THELMA TODD'S CAFE
MIRAMAR HOTEL
DEL MAR CLUB
NORMA SHEARER
DEAUVILLE CLUB

Hollywood starland: official moviegraph of the land of stars, where they live, where they work and where they play, by Don Boggs (1937).

For many people life is often tedious, so they seek relief in fiction. Americans do this most often by going to the movies, where they may very well identify not just with the fictional characters but also with the stars and starlets who create them on film. Some regard big movie stars to be royalty, and in California they are treated that way. As early as the 1930s, guides in Los Angeles were offering tours with stops outside the homes of well-known film personalities. This map of Hollywood's "Starland" was sold during the late 1930s to tourists who wanted a do-it-yourself tour.

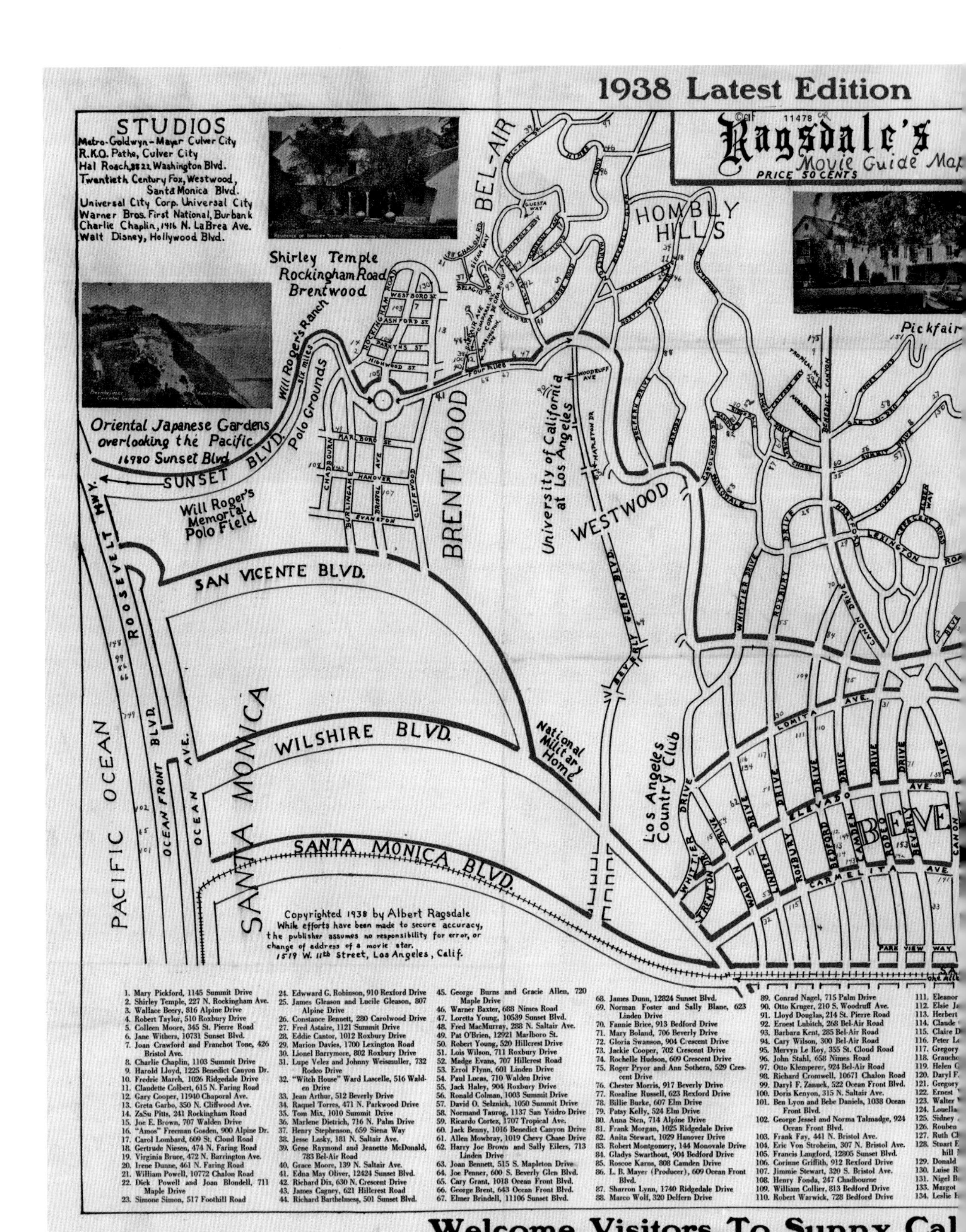

1938 Latest Edition
Ragsdale's
Movie Guide Map
PRICE 50 CENTS
STUDIOS
Metro-Goldwyn-Mayer Culver City
R.K.O. Pathe, Culver City
Hal Roach, 8822 Washington Blvd.
Twentieth Century Fox, Westwood, Santa Monica Blvd.
Universal City Corp. Universal City
Warner Bros. First National, Burbank
Charlie Chaplin, 1416 N. LaBrea Ave.
Walt Disney, Hollywood Blvd.
Shirley Temple Rockingham Road Brentwood
Oriental Japanese Gardens overlooking the Pacific 16980 Sunset Blvd.
BEL-AIR
HOMBLY HILLS
Pickfair
Will Roger's Ranch six miles
Polo Grounds
SUNSET BLVD.
Will Roger's Memorial Polo Field
BRENTWOOD
University of California at Los Angeles
WESTWOOD
SAN VICENTE BLVD.
ROOSEVELT HWY.
PACIFIC OCEAN
OCEAN FRONT BLVD.
OCEAN AVE.
SANTA MONICA
WILSHIRE BLVD.
National Military Home
Los Angeles Country Club
SANTA MONICA BLVD.
BEVERLY GLEN BLVD.
Copyrighted 1938 by Albert Ragsdale
While efforts have been made to secure accuracy, the publisher assumes no responsibility for error, or change of address of a movie star.
1519 W. 11th Street, Los Angeles, Calif.
1. Mary Pickford, 1145 Summit Drive
2. Shirley Temple, 227 N. Rockingham Ave.
3. Wallace Beery, 816 Alpine Drive
4. Robert Taylor, 510 Roxbury Drive
5. Colleen Moore, 345 St. Pierre Road
6. Jane Withers, 10731 Sunset Blvd.
7. Joan Crawford and Franchot Tone, 426 Bristol Ave.
8. Charlie Chaplin, 1103 Summit Drive
9. Harold Lloyd, 1225 Benedict Canyon Dr.
10. Fredric March, 1026 Ridgedale Drive
11. Claudette Colbert, 615 N. Faring Road
12. Gary Cooper, 11940 Chaporal Ave.
13. Greta Garbo, 350 N. Cliffwood Ave.
14. ZaSu Pitts, 241 Rockingham Road
15. Joe E. Brown, 707 Walden Drive
16. "Amos" Freeman Gosden, 900 Alpine Dr.
17. Carol Lombard, 609 St. Cloud Road
18. Gertrude Niesen, 474 N. Faring Road
19. Virginia Bruce, 472 N. Barrington Ave.
20. Irene Dunne, 461 N. Faring Road
21. William Powell, 10772 Chalon Road
22. Dick Powell and Joan Blondell, 711 Maple Drive
23. Simone Simon, 517 Foothill Road
24. Edwward G. Robinson, 910 Rexford Drive
25. James Gleason and Lucile Gleason, 807 Alpine Drive
26. Constance Bennett, 280 Carolwood Drive
27. Fred Astaire, 1121 Summit Drive
28. Eddie Cantor, 1012 Roxbury Drive
29. Marion Davies, 1700 Lexington Road
30. Lionel Barrymore, 802 Roxbury Drive
31. Lupe Velez and Johnny Weismuller, 732 Rodeo Drive
32. "Witch House" Ward Lascelle, 516 Walden Drive
33. Jean Arthur, 512 Beverly Drive
34. Raquel Torres, 471 N. Parkwood Drive
35. Tom Mix, 1010 Summit Drive
36. Marlene Dietrich, 716 N. Palm Drive
37. Henry Stephenson, 659 Siena Way
38. Jesse Lasky, 181 N. Saltair Ave.
39. Gene Raymond and Jeanette McDonald, 783 Bel-Air Road
40. Grace Moore, 139 N. Saltair Ave.
41. Edna May Oliver, 12424 Sunset Blvd.
42. Richard Dix, 630 N. Crescent Drive
43. James Cagney, 621 Hillcrest Road
44. Richard Barthelmess, 501 Sunset Blvd.
45. George Burns and Gracie Allen, 720 Maple Drive
46. Warner Baxter, 688 Nimes Road
47. Loretta Young, 10539 Sunset Blvd.
48. Fred MacMurray, 288 N. Saltair Ave.
49. Pat O'Brien, 12921 Marlboro St.
50. Robert Young, 520 Hillcrest Drive
51. Lois Wilson, 711 Roxbury Drive
52. Madge Evans, 707 Hillcrest Road
53. Errol Flynn, 601 Linden Drive
54. Paul Lucas, 710 Walden Drive
55. Jack Haley, 904 Roxbury Drive
56. Ronald Colman, 1003 Summit Drive
57. David O. Selznick, 1050 Summit Drive
58. Normand Taurog, 1137 San Ysidro Drive
59. Ricardo Cortez, 1707 Tropical Ave.
60. Jack Benny, 1016 Benedict Canyon Drive
61. Allen Mowbray, 1019 Chevy Chase Drive
62. Harry Joe Brown and Sally Eilers, 713 Linden Drive
63. Joan Bennett, 515 S. Mapleton Drive
64. Joe Penner, 600 S. Beverly Glen Blvd.
65. Cary Grant, 1018 Ocean Front Blvd.
66. George Brent, 643 Ocean Front Blvd.
67. Elmer Brindell, 11106 Sunset Blvd.
68. James Dunn, 12824 Sunset Blvd.
69. Norman Foster and Sally Blane, 623 Linden Drive
70. Fannie Brice, 913 Bedford Drive
71. Mary Boland, 706 Beverly Drive
72. Gloria Swanson, 904 Crescent Drive
73. Jackie Cooper, 702 Crescent Drive
74. Rochelle Hudson, 609 Crescent Drive
75. Roger Pryor and Ann Sothern, 529 Crescent Drive
76. Chester Morris, 917 Beverly Drive
77. Rosaline Russell, 623 Rexford Drive
78. Billie Burke, 607 Elm Drive
79. Patsy Kelly, 524 Elm Drive
80. Anna Sten, 714 Alpine Drive
81. Frank Morgan, 1025 Ridgedale Drive
82. Anita Stewart, 1029 Hanover Drive
83. Robert Montgomery, 144 Monovale Drive
84. Gladys Swarthout, 904 Bedford Drive
85. Roscoe Karns, 808 Camden Drive
86. L. B. Mayer (Producer), 609 Ocean Front Blvd.
87. Sharron Lynn, 1740 Ridgedale Drive
88. Marco Wolf, 320 Delfern Drive
89. Conrad Nagel, 715 Palm Drive
90. Otto Kruger, 210 S. Woodruff Ave.
91. Lloyd Douglas, 214 St. Pierre Road
92. Ernest Lubitch, 268 Bel-Air Road
93. Barbara Kent, 285 Bel-Air Road
94. Cary Wilson, 300 Bel-Air Road
95. Mervyn Le Roy, 355 St. Cloud Road
96. John Stahl, 658 Nimes Road
97. Otto Klemperer, 924 Bel-Air Road
98. Richard Cromwell, 10671 Chalon Road
99. Daryl F. Zanuck, 522 Ocean Front Blvd.
100. Doris Kenyon, 315 N. Saltair Ave.
101. Ben Lyon and Bebe Daniels, 1038 Ocean Front Blvd.
102. George Jessel and Norma Talmadge, 924 Ocean Front Blvd.
103. Frank Fay, 441 N. Bristol Ave.
104. Eric Von Stroheim, 307 N. Bristol Ave.
105. Francis Langford, 12805 Sunset Blvd.
106. Corinne Griffith, 912 Rexford Drive
107. Jimmie Stewart, 320 S. Bristol Ave.
108. Henry Fonda, 247 Chadbourne
109. William Collier, 813 Bedford Drive
110. Robert Warwick, 728 Bedford Drive
Welcome Visitors To Sunny Cal

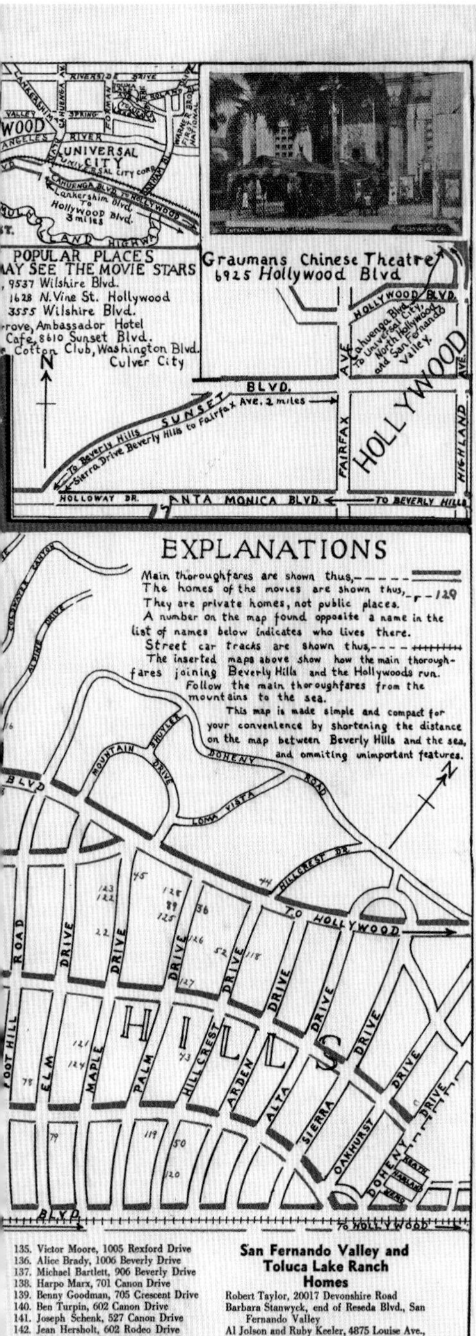

135. Victor Moore, 1005 Rexford Drive
136. Alice Brady, 1006 Beverly Drive
137. Michael Bartlett, 906 Beverly Drive
138. Harpo Marx, 701 Canon Drive
139. Benny Goodman, 705 Crescent Drive
140. Ben Turpin, 602 Canon Drive
141. Joseph Schenk, 527 Canon Drive
142. Jean Hersholt, 602 Rodeo Drive
143. Walter Connolly, 601 Camden Drive
144. Dixie Dunbar, 609 Camden Drive
145. Carl Laemmle, Sr., 1251 Benedict Canyon
146. Buddy Desylvia, 444 Faring Road
147. Betty Grable, 250 Chadbourne
148. Marion Davies, 415 Ocean Front Blvd.
149. Norma Shearer, 707 Ocean Front Blvd.
150. King Vidor, 1120 Summit Drive
151. Miriam Hopkins, Tower Road
152. Fred Stone, 1020 Ridgedale Drive
153. Nelson Eddy, 517 N. Beverly Drive
154. Myrna Loy, Arthur Hornblow, Jr., Hidden Valley Rd., out Coldwater Canyon.
155. Mary Pickford, Buddy Rogers, 426 Rockingham Ave.

San Fernando Valley and Toluca Lake Ranch Homes

Robert Taylor, 20017 Devonshire Road
Barbara Stanwyck, end of Reseda Blvd., San Fernando Valley
Al Jolson and Ruby Keeler, 4875 Louise Ave., San Fernando Valley
Edward Everett Horton, 5521 Amestoy Ave.
Warren Williams, 4717 Encino Ave.
Abner (Lum and Abner Radio Team), 4849 Encino Ave.
Spencer Tracy, 5508 White Oak Ave.
Robert Armstrong, 5337 White Oak Ave.
Paul Kelly, 20005 Devonshire Road
Frank Buck (Bring 'em back alive), 5035 Louise Ave.
Phil Harris (orchestra leader), 4544 Encino Ave.
Sterling Holloway, 4510 Encino Ave.
Benie Rubin, 5439 Encino Ave.
Richard Arlen, 10025 Toluca Lake Ave.
Bing Crosby, 10500 Camarillo

Ragsdale's movie guide map, by Albert Ragsdale (1938).

One of the most widely known and admired neighborhoods in the world, Beverly Hills has served as an enclave for generations of wealthy producers, directors, and actors. Big stars started buying lots in Beverly Hills after Douglas Fairbanks and Mary Pickford built a home there in 1921. John Barrymore, Miriam Hopkins, and scores of other film personalities soon followed. Before long, tourists started making their own appearances in Beverly Hills. The 1938 Ragsdale map was intended as a guide for visitors who hoped to catch a glimpse of the rich and famous.

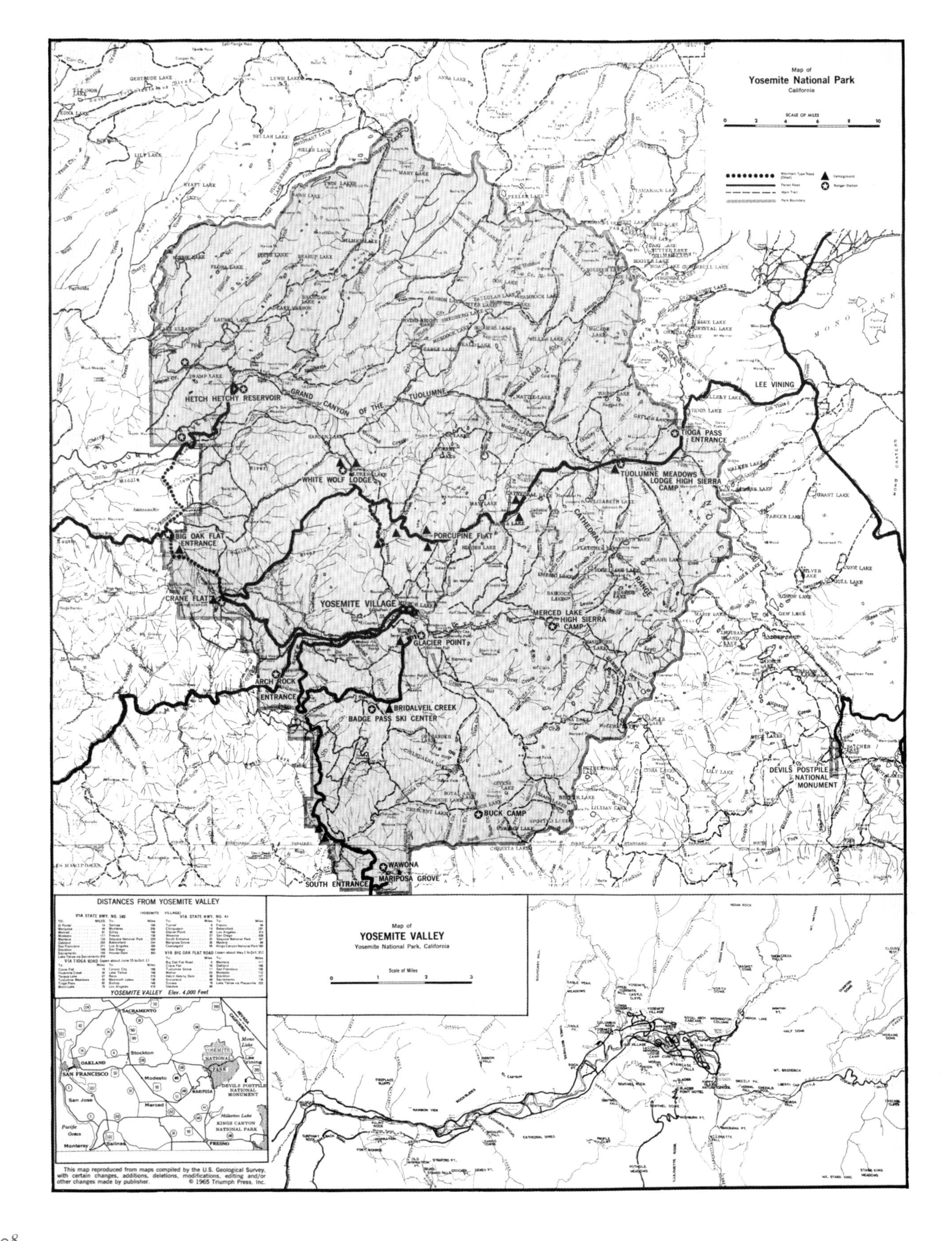

Map of
Yosemite National Park
California
SCALE OF MILES
HETCH HETCHY RESERVOIR
GRAND CANYON OF THE TUOLUMNE
WHITE WOLF LODGE
TIOGA PASS ENTRANCE
TUOLUMNE MEADOWS LODGE HIGH SIERRA CAMP
LEE VINING
MONO LAKE
PORCUPINE FLAT
BIG OAK FLAT ENTRANCE
CRANE FLAT
YOSEMITE VILLAGE
CATHEDRAL RANGE
MERCED LAKE HIGH SIERRA CAMP
GLACIER POINT
ARCH ROCK ENTRANCE
BRIDALVEIL CREEK
BADGE PASS SKI CENTER
DEVILS POSTPILE NATIONAL MONUMENT
BUCK CAMP
WAWONA
MARIPOSA GROVE
SOUTH ENTRANCE
DISTANCES FROM YOSEMITE VALLEY
YOSEMITE VALLEY Elev. 4,000 Feet
Map of
YOSEMITE VALLEY
Yosemite National Park, California
Scale of Miles
SACRAMENTO
Stockton
OAKLAND
SAN FRANCISCO
Modesto
San Jose
Merced
Monterey
Salinas
FRESNO
YOSEMITE NATIONAL PARK
DEVILS POSTPILE NATIONAL MONUMENT
KINGS CANYON NATIONAL PARK
This map reproduced from maps compiled by the U.S. Geological Survey, with certain changes, additions, deletions, modifications, editing and/or other changes made by publisher. © 1965 Triumph Press, Inc.

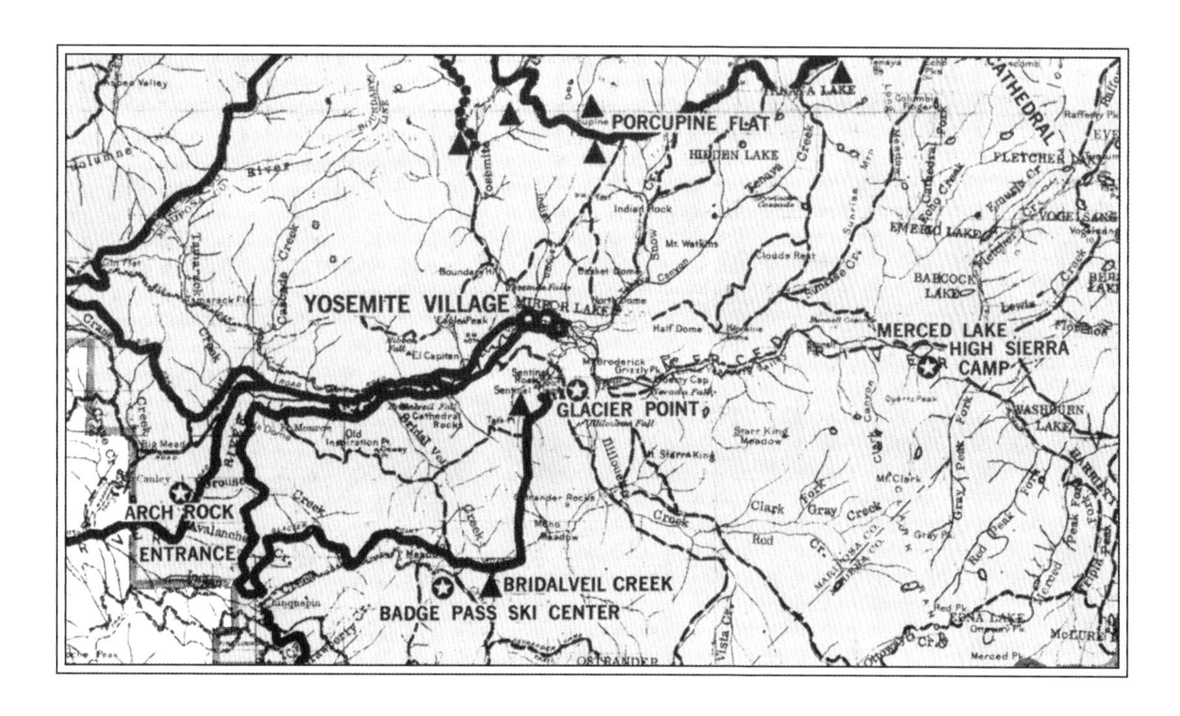

Map of Yosemite National Park, U.S. Geological Survey (1965).

Deeply ingrained in Californians is the notion that nature is the best medicine for whatever ails the human body and spirit. A visit to Yosemite, Big Sur, or the 300-foot-tall coastal sequoia forest may convince nearly anyone that this is so. Unfortunately, as more than 30 million people have crowded into the state over the last century, the need to house, feed, and employ them has threatened the very natural wonders that attracted many of them in the first place. It is all but certain that without federal protection, the finest features of Yosemite would have been destroyed long ago. Indeed, Hetchy Hetch, a similarly spectacular canyon just north of Yosemite, was inundated by a reservoir in 1923 to provide water for San Francisco. This USGS map shows 1,200-square-mile Yosemite National Park.

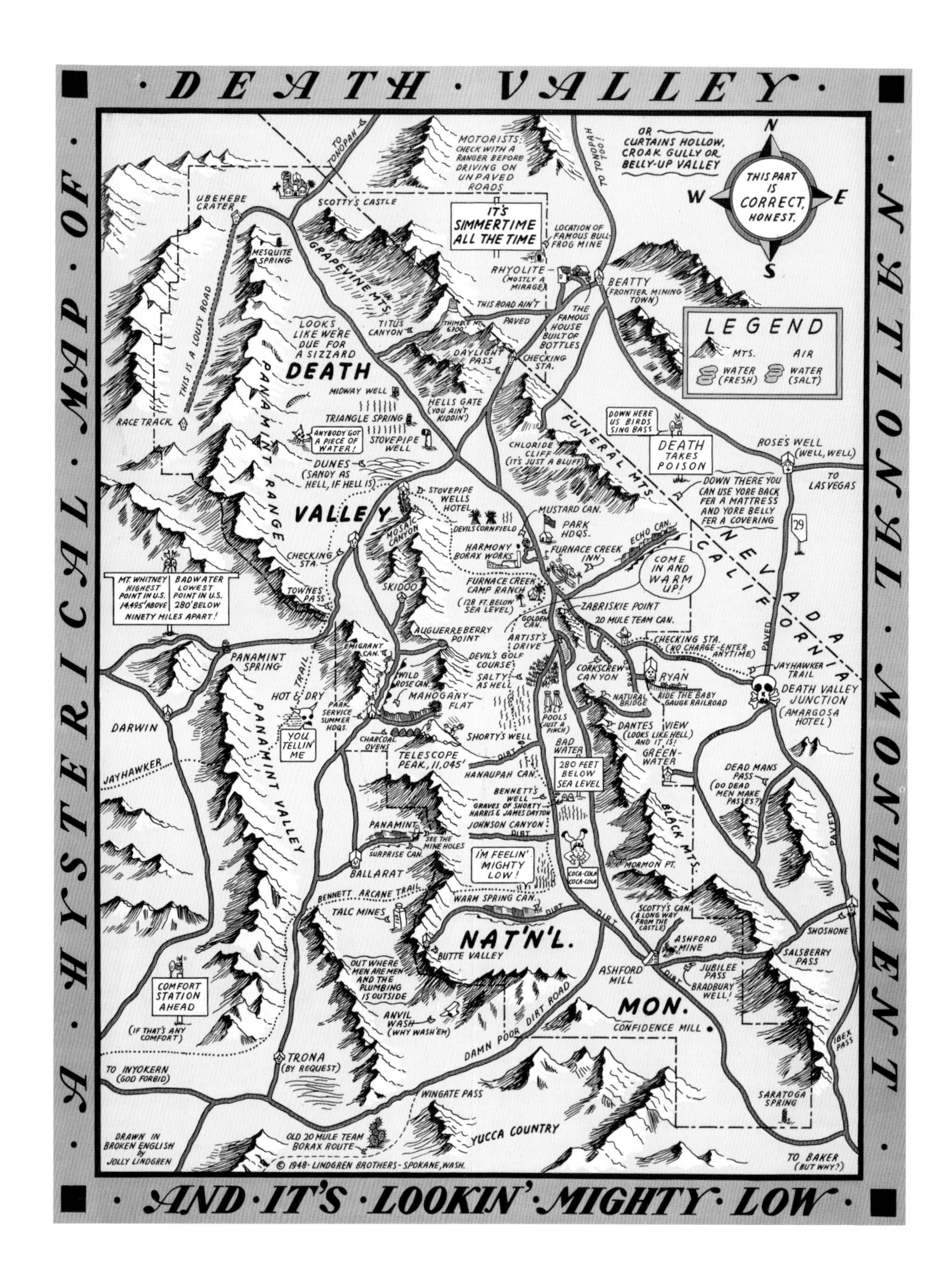

A HYSTERICAL MAP OF DEATH VALLEY NATIONAL MONUMENT
AND IT'S LOOKIN' MIGHTY LOW
OR CURTAINS HOLLOW, CROAK GULLY OR BELLY-UP VALLEY
THIS PART IS CORRECT, HONEST.
MOTORISTS: CHECK WITH A RANGER BEFORE DRIVING ON UNPAVED ROADS
IT'S SIMMERTIME ALL THE TIME
LEGEND
MTS.
AIR
WATER (FRESH)
WATER (SALT)
DEATH
VALLEY
NAT'N'L.
MON.
TO TONOPAH
TO TONOPAH TOO!
UBEHEBE CRATER
SCOTTY'S CASTLE
LOCATION OF FAMOUS BULL-FROG MINE
MESQUITE SPRING
GRAPEVINE MTS.
RHYOLITE (MOSTLY A MIRAGE)
BEATTY (FRONTIER MINING TOWN)
THIS ROAD AIN'T PAVED
THE FAMOUS HOUSE BUILT OF BOTTLES
THIS IS A LOUSY ROAD
LOOKS LIKE WE'RE DUE FOR A SIZZARD
TITUS CANYON
THIMBLE MT. 6300'
DAYLIGHT PASS
CHECKING STA.
PANAMINT RANGE
MIDWAY WELL
HELLS GATE (YOU AIN'T KIDDIN')
TRIANGLE SPRING
RACE TRACK
ANYBODY GOT A PIECE OF WATER?
STOVEPIPE WELL
DOWN HERE US BIRDS SING BASS
FUNERAL MTS.
CHLORIDE CLIFF (IT'S JUST A BLUFF)
DEATH TAKES POISON
ROSE'S WELL (WELL, WELL)
DUNES (SANDY AS HELL, IF HELL IS)
STOVEPIPE WELLS HOTEL
DOWN THERE YOU CAN USE YORE BACK FER A MATTRESS AND YORE BELLY FER A COVERING
TO LAS VEGAS
MUSTARD CAN.
PARK HDQS.
DEVILS CORNFIELD
MOSAIC CANYON
HARMONY BORAX WORKS
FURNACE CREEK INN
ECHO CAN.
NEVADA
CALIFORNIA
29
CHECKING STA.
COME IN AND WARM UP!
MT. WHITNEY HIGHEST POINT IN U.S. 14,495' ABOVE
BADWATER LOWEST POINT IN U.S. 280' BELOW
NINETY MILES APART!
TOWNES PASS
SKIDOO
FURNACE CREEK CAMP RANCH (128 FT. BELOW SEA LEVEL)
ZABRISKIE POINT
20 MULE TEAM CAN.
GOLDEN CAN.
AUGUERREBERRY POINT
ARTIST'S DRIVE
PAVED
CHECKING STA. (NO CHARGE - ENTER ANYTIME)
PANAMINT SPRING
EMIGRANT CAN.
TRAIL
DEVIL'S GOLF COURSE
SALTY AS HELL
CORKSCREW CANYON
RYAN
JAYHAWKER TRAIL
WILD ROSE CAN.
HOT & DRY
MAHOGANY FLAT
NATURAL BRIDGE
RIDE THE BABY GAUGE RAILROAD
DEATH VALLEY JUNCTION (AMARGOSA HOTEL)
PARK SERVICE SUMMER HDQS.
SALT POOLS JUST A PINCH
DARWIN
PANAMINT VALLEY
YOU TELLIN' ME
DANTES VIEW (LOOKS LIKE HELL) AND IT IS!
CHARCOAL OVENS
SHORTY'S WELL
BAD WATER
GREEN-WATER
TELESCOPE PEAK, 11,045'
DIRT
JAYHAWKER
280 FEET BELOW SEA LEVEL
HANAUPAH CAN.
DEAD MANS PASS (DO DEAD MEN MAKE PASSES?)
BENNETT'S WELL
GRAVES OF SHORTY HARRIS & JAMES DAYTON
BLACK MTS.
PANAMINT
SEE THE MINE HOLES
JOHNSON CANYON
SURPRISE CAN.
I'M FEELIN' MIGHTY LOW!
BALLARAT
COCA-COLA
MORMON PT.
PAVED
BENNETT ARCANE TRAIL
WARM SPRING CAN.
TALC MINES
SCOTTY'S CAN. (A LONG WAY FROM THE CASTLE)
ASHFORD MINE
SHOSHONE
BUTTE VALLEY
SALSBERRY PASS
COMFORT STATION AHEAD
OUT WHERE MEN ARE MEN AND THE PLUMBING IS OUTSIDE
ASHFORD MILL
JUBILEE PASS
BRADBURY WELL!
(IF THAT'S ANY COMFORT)
ANVIL WASH (WHY WASH 'EM)
DAMN POOR DIRT ROAD
CONFIDENCE MILL
IBEX PASS
TRONA (BY REQUEST)
TO INYOKERN (GOD FORBID)
WINGATE PASS
SARATOGA SPRING
YUCCA COUNTRY
DRAWN IN BROKEN ENGLISH by JOLLY LINDGREN
OLD 20 MULE TEAM BORAX ROUTE
© 1948 · LINDGREN BROTHERS · SPOKANE, WASH.
TO BAKER (BUT WHY?)

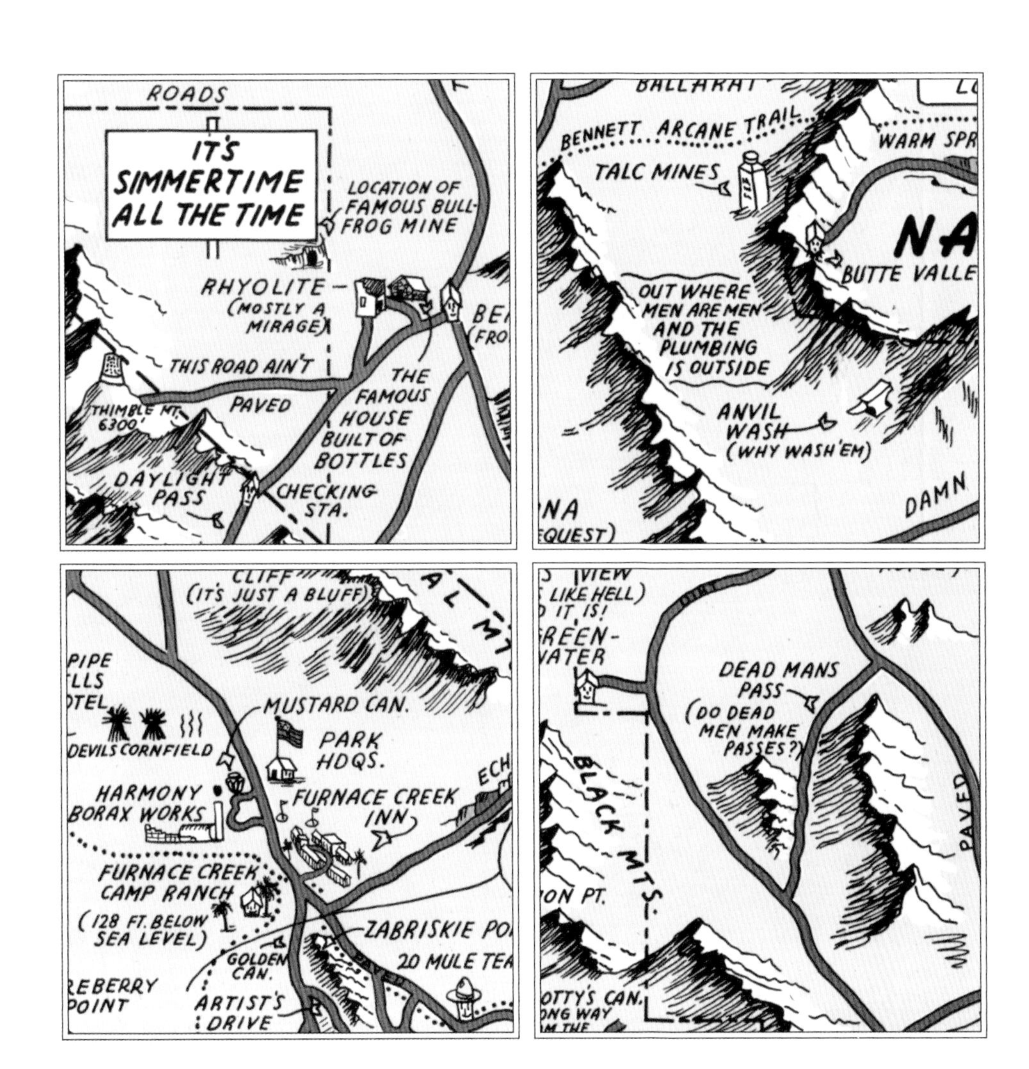

Pictorial Map of Death Valley, by Jolly Lindgren (1948).

To pioneers and prospectors trying to cross it during the nineteenth century, Death Valley may have looked harmless enough and far less challenging than the towering Sierras beyond. However, some would never live long enough to take on the mountains. Dipping down some 280 feet below sea level, the valley floor acts as the sun's anvil, driving temperatures above 130 degrees. Years can pass between rains in Death Valley. Forbidding though it may be, the valley is extraordinarily beautiful, as visitors to Death Valley National Park soon learn. A Death Valley map produced in 1948 for sale to park visitors points out valley attractions.

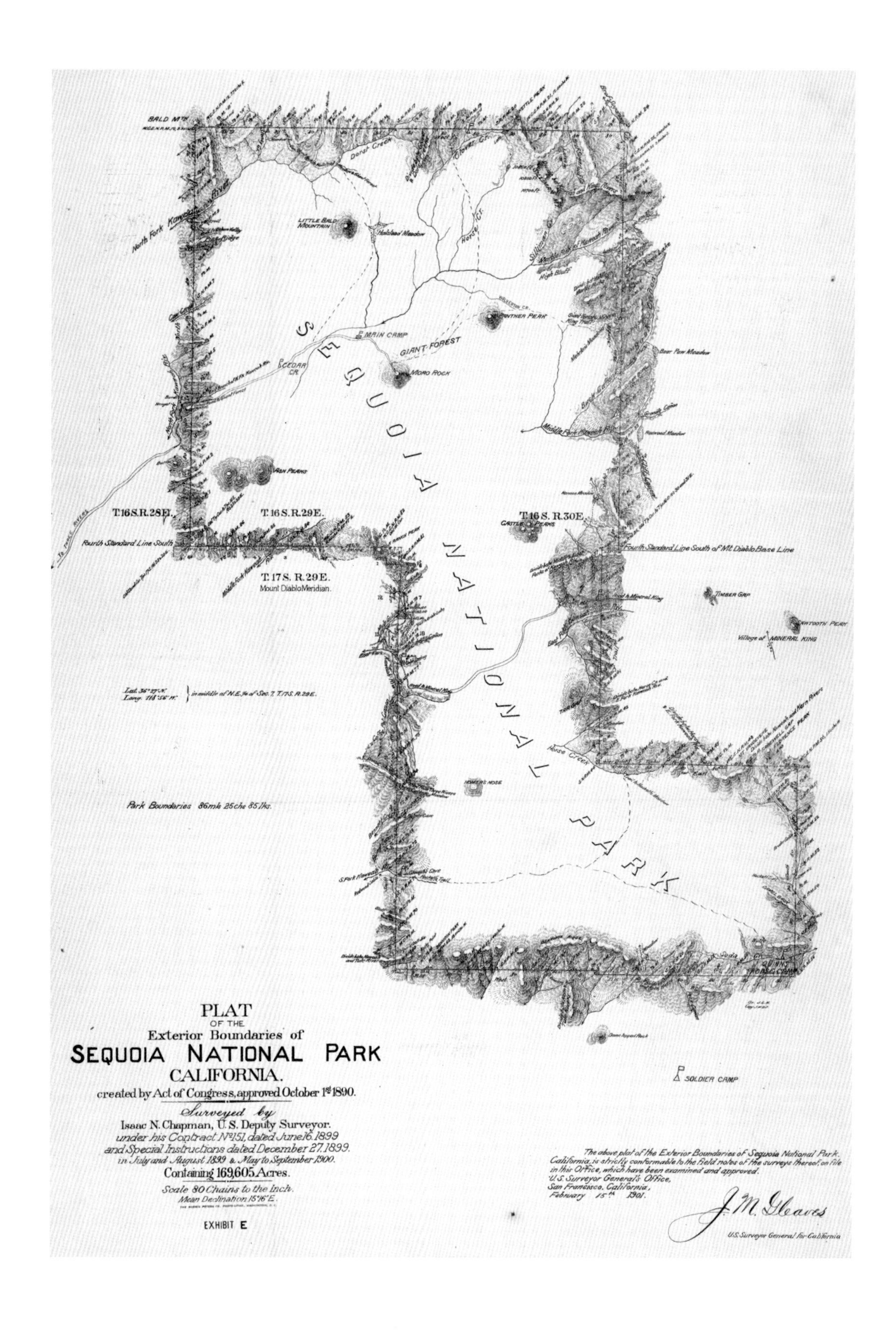

BALD MTN
Dorst Creek
LITTLE BALD MOUNTAIN
North Fork Kaweah River
PANTHER PEAK
MAIN CAMP
GIANT FOREST
CEDAR CR.
MORO ROCK
SEQUOIA NATIONAL PARK
Bear Paw Meadow
T.16S.R.28E.
T.16S.R.29E.
T.16S.R.30E.
CASTLE PEAKS
Fourth Standard Line South
Fourth Standard Line South of Mt. Diablo Base Line
T.17S. R.29E.
Mount Diablo Meridian.
TIMBER GAP
SAWTOOTH PEAK
Village of MINERAL KING
Park Boundaries 86mls 25chs 85lks.
Rose Creek
HOMER'S NOSE
SOLDIER CAMP
PLAT
OF THE
Exterior Boundaries of
SEQUOIA NATIONAL PARK
CALIFORNIA.
created by Act of Congress, approved October 1st 1890.
Surveyed by
Isaac N. Chapman, U.S. Deputy Surveyor.
under his Contract No. 151, dated June 16, 1899
and Special Instructions dated December 27, 1899.
in July and August 1899 & May to September 1900.
Containing 169,605 Acres.
Scale 80 Chains to the Inch.
Mean Declination 15°16' E.
EXHIBIT E
The above plat of the Exterior Boundaries of Sequoia National Park, California, is strictly conformable to the field notes of the surveys thereof on file in this Office, which have been examined and approved.
U.S. Surveyor General's Office,
San Francisco, California,
February 15th 1901.
J. M. Gleaves
U.S. Surveyor General for California

Conclusion

What is California? California is a strip of land approximately 800 miles long and 250 miles wide along the southwestern and central Pacific coasts of the United States. It spreads across 158,707 square miles, or about 100 million acres. This makes it the third largest state after Alaska and Texas. If sticking reasonably close to the freeway speed limits—and in California few drivers do—it can take seventeen to eighteen hours to drive from the Mexican border in the south to the Oregon border in the north.

What is California? California is a community of many millions. Because it is growing so fast, California's population is difficult to measure at any one point in time, but by 2010 it is expected to surpass 40 million. This makes California the largest of the fifty states in terms of population. California's population is six or seven times that of an average-sized state such as Missouri (about 6 million) and more than fifty times that of states with small populations such as North Dakota (less than 700,000). California has more than a dozen cities with metropolitan populations larger than that of the entire state of Wyoming (about 500,000). And California's population is growing very fast. Nearly

Plat of the Exterior Boundaries of Sequoia National Park, Isaac N. Chapman (1901).
A plat shows the boundaries of Sequoia National Park, founded in 1890 to protect the last remaining groves of giant sequoias. These extraordinary trees, belonging to a subspecies related to the coastal redwood, can live as long as thirty-five hundred years and reach a height of more than 280 feet and a diameter of more than 50 feet. Incredibly, until the park was established, these ancient trees were cut for timber, usually at a loss for the lumbering operations involved. Since it splintered easily, the wood proved nearly useless for construction.

every year California adds to its population more than enough people to fill a city the size of Atlanta or Cincinnati or a state the size of Vermont.

California's people are extraordinarily diverse. There are Californians who were born in every one of the world's two hundred or so nations. Californians speak a polyglot of more than one hundred different languages; in fact, only about 60 percent of Californians speak English at home. Another 25 percent speak Spanish, while the rest speak Chinese, Japanese, Vietnamese, Filipino, Russian, Italian, or any one of numerous other languages. Even some of the Californians who commonly communicate in English do so in a dialect or style that is alien to Americans from the East or Midwest. Certain forms of speech, such as the widely satirized Valley Speak—"don't you know"—are California creations.

It is common for Californians to ask new acquaintances where they are from because very often that place is somewhere else. The state is home to surprisingly large ethnic communities that continue to celebrate their own rich heritages. For instance, California has more than six hundred thousand Armenian Americans, half a million Iranian Americans, and about that same number of Arab Americans. There are at least one hundred thousand Californians from the Pacific Islands, including some twenty-five thousand from Samoa. Of course, there are also very large Asian-American communities, including Chinese, Japanese, and Indian. Even the small inland nation of Laos is well represented. There are nearly nine thousand Laotians in the city of Fresno alone.

Five centuries ago the California population was made up entirely of indigenous peoples, and today the state continues to harbor a very large Native American population. Every North American tribe is represented to some extent in California. The state's total Native American population approaches four hundred thousand. Cherokees, many of whom migrated here from Oklahoma during the Dust Bowl days of the Great Depression, comprise the largest single group of Native Americans in California, with a statewide tribal population of about one hundred thousand.

California is about as balanced as any place could be when it comes to religion. The state is nearly evenly divided between Roman Catholics, Protestants, and people who practice other religions or no religion at all. California's Jewish synagogues have about a million members. There are also churches, temples, and centers devoted to Buddhism, Hinduism, Shinto, Sikhism, Taoism, and countless other forms of worship. There are Tibetan Buddhist monasteries in the mountains and many other monastic cloisters of one sort or another scattered around the state. And of course, Californians are not above inventing their own religions. It is always possible to find Californians who have discovered a technique for "channeling" themselves into another dimension. Or perhaps they have come from that dimension.

So how do all these diverse millions manage to live together in relative peace and harmony? They manage it quite well, thank you very much. They treat one another with respect and understanding because they know that all Californians, whatever their religions or ethnic backgrounds, have one very important thing in common. They each cling to their own version of the California dream. People here have had this in common throughout California's long and fascinating history. They are citizens of a state where anything and everything is believed to be possible. If their own dreams can be fulfilled, then they must allow for the possibility that the dreams of their neighbors may be fulfilled as well.

Acknowledgments

The publisher and the authors gratefully acknowledge the staff at the Library of Congress for their fine work and research assistance on this book, particularly Colleen Cahill, Aimee Hess, and Ralph Eubanks.

Without the vision and professionalism of Erin Turner at Globe Pequot Press, this audacious project would not be the permanent achievement it is bound to be.

—Vincent Virga

My sincerest thanks to my good friend and fellow Globe Pequot Press author Joe Lubow for his invaluable research assistance. I would also like to thank former Globe editor Amy Paradysz for her encouragement and her faith in me as an author. It was Amy's hard work and boundless enthusiasm that made this series possible. Finally, my thanks go out to the multitalented Vincent Virga and others at the Library of Congress who contributed mightily to this book.

—Ray Jones

Notes

All maps come from the Library of Congress Geography and Map Division unless otherwise noted. To order reproductions of Library of Congress items, please contact the Library of Congress Photoduplication Service, Washington, D.C. 20540-4570 or (202) 707-5640.

Pages ii and 6–7 Gutiérrez Map of America. Diego Gutiérrez, *Americae sive Quartae Orbis Partis Nova et Exactissima Descriptio*, Antwerp, 1562. G3290 1562. G7. Vault oversize.

Page vi Ruysch, Johann. "Universalior cogniti orbis tabula." In Claudius Ptolemeus, *Geographia.* Rome, 1507. G1005.1507 Vault

Page vii Waldseemüller, Martin. "Universalis cosmographia secudum Ptholomaei traditionem et Americi Vespucii aloru[m] que lustrations," St. Dié, France?, 1507. G32001507.W3 Vault

Page 2 Mexico (West Coast), 1535.

Pages 8–9 *La herdike enterprinse faict par le Signeur Draeck D'Avoir cirquit toute la Terre.* Nicola van Sype. Antwerp: s.n., 1581. G3201.S12 1581 .S9

Pages 10–11 Map of California shown as an island. Vinckeboons, Joan. [ca. 1650] G3291.S12 coll .H3 Vault: Harr vol. 2, map. 10

Page 12 Descriptión de las Yndias Ocidentalis. In Antonio de Herrera y Tordesillas. Description des Indes Occidentales. Amsterdam: M. Colin, 1622. Rare Book and Special Collections Division. Thacher A754 Thacher Coll

Page 13 "Via Terrestis in California . . . Anno 1698 ad annum 1701" in Der Neue Welt-Bott . . . Vol. 1, pt. 2. Augspurg und Grätz: 1726–1758. Engraved map. Rare Book and Special Collections Division. BV2290.A27 1726 (office)

Pages 14–15 La Californie ou Nouvelle Caroline: teatro de los trabajos, Apostolicos de la Compa. e Jesus en la America Septe. / par N. de Fer, Geographe de sa Majesté Catolique. Fer, Nicolas de, 1646–1720. Paris: [N. de Fer], 1720. G4410 1720 .F42 TIL Vault

Pages 16 and 20–21 *Carte de l'Amérique septentrionale,* 1754. Palairet, Jean, 1697–1774. [Londres, 1755] G3300 1754 .P3 Vault

Pages 22–23 California missions. Newman, Wm. L. Glendale, Calif.: Wm. L. Newman, 1949. G4361.E424 1949 .N4

Pages 24–25 The old Spanish and Mexican ranchos of Los Angeles County, Gerald A. Eddy, Los Angeles: Title Insurance and Trust Company, Los Angeles; c1937 G4363.L6 1937 .E3

Page 26 Louisiana. Lewis, Samuel, 1753 or 4-1822. [S.l., 1805]. From Arrowsmith & Lewis New and Elegant General Atlas, 1804. G4050 1805 .L4 TIL

Page 27 A map of the Internal Provinces of New Spain. Pike, Zebulon Montgomery, 1779–1813. [S.l., 1807]. G4295 1807 .P5 TIL

Pages 28–29 Missouri territory, formerly Louisiana. Carey, Mathew, 1760–1839. [S.l., 1814] G4050 1814 .C3 TIL

Pages 30–31 Important Historical Events Which Have Made Los Angeles' Growth Possible, Gerald A. Eddy. [S.l.]: Frank L. Meline, Inc., c1929. G4361.S1 1929 .E4

Pages 32, 36–37 Map of Oregon and upper California from the surveys of John Charles Frémont and other authorities, drawn by Charles Preuss under the order of the Senate of the United States; lithy. by E. Weber & Co., Balto. Frémont, John Charles, 1813–1890. Washington, D.C.: The Senate, 1848. G4210 1848 .F72 Vault: Fil 136

Page 38–39 Map of the United States, the British provinces, Mexico &c. Atwood, John M., b. ca. 1818. New York: J. H. Colton, 1849. G3700 1849 .A72 TIL

Page 40 Map of the mining district of California. Jackson, Wm. A. (William A.) [S.l., 1851]. G4361.H1 1851 .J3 TIL

Page 41 Map of passes in the Sierra Nevada from Walker's Pass to the Coast Range: from explorations and surveys, made under the direction of the Hon. Jefferson Davis, Secretary of War by Lieut. R. S. Williamson Topl. Engr. assisted by Lieut. J. G. Parke Topl. Engr. and Mr. Isaac Williams Smith, Civ. Engr., 1853. Williamson, R. S. 1824–1882. (Robert Stockton) [Washington, D.C., 1859], G4360 1853 .W51 RR 150

Pages 42–43 View of the Panamint Range Mountains, mines, mills and town site; Sherman Town, property of the Panamint Mining & Concentration Works. Britton, Rey & Co. [S.l., 1875]. G4362.P213H1 1875 .B7 TIL

Pages 44–45 Views of oil fields around Los Angeles / C. S. & E. M. Forncrook. 1922 G4361.H8 1922 .C2

Pages 46–47 Official map of the County of Napa, California: compiled from the official records and latest surveys, by O. H. Buckman. San Francisco Punnett Bros. 1895. G4363.N3 1895 .B81

Page 48 and 52 Map of California to accompany printed agreement of S. O. Houghton as to the rights of the Southern Pacific R.R. Co. of Cal. to government lands under Acts of Congress passed July 27, 1866, and March 3, 1871, made before the committee of the judiciary of the Senate and Ho. of Reps. in May 1876. G. W. & C. B. Colton & Co. [n.p., 1876] G4361.P3 1876 .G15 RR 568

Page 53 New enlarged scale railroad and county map of California showing every railroad station and post office in the state. Rand McNally and Company. Chicago, c1883. G4361.P3 1883 .R3 RR 189

Pages 54–55 Official map of the County of Solano, California: showing Mexican grants, United States government and swamp land surveys, present private land ownerships, roads and railroads, Compiled by E. N. Eager, County Surveyor; approved by the Board of Supervisors. Eager, E. N. (Edward Nelson) April 7, 1890. [California: Solano County], 1890 (S[an] F[rancisco]: Britton & Rey) G4363.S7 1890 .E2

Pages 56–57 Map of California roads for cyclers. Blum, George W. [S.l., 1895]. G4361.P2 1895 .B5 TIL

Page 58 Automobile routes from Los Angeles to Sunland, La Crescenta, and La Cañada, Touring bureau. Route and map service. Automobile Club of Southern California. 1912. Los Angeles: Automobile Club of Southern California, 1912. G4360 1912 .A8

Pages 62–63 Map of San Francisco, showing principal streets and places of interest, Harrison Godwin, 1927. Ethel Fair collection

Pages 64–65 Los Angeles as it appeared in 1871. *Gores, fecit.* [Los Angeles] Women's University Club of L.A., 1929. G4364.L8A3 1871 .G6

Pages 66–67 Auburn, Cal., C. P. Cook., del.; presented with the compliments of W. B. Lardner, real estate agent, att'y-at-law & notary public. Cook, C. P. S[an] F[rancisco]: W. W. Elliott [1887?], G4364.A86A3 1887 .C6

Pages 68–69 Bird's eye view of Azusa, Los Angeles Co. Cal., 1887, E. S. Moore, del. [Los Angeles, Calif.?: Slauson], 1887. G4364.A9A3 1887 .M61

Pages 70–71 Bird's eye view of Coronado Beach, San Diego Bay and city of San Diego, Cal. in distance, sketch by E. S. Moore ; Crocker & Co., lith., S.F. Moore, E. S. [San Diego]: Coronado Beach Co., [188-] G4364.C77A3 188- .M6 MLC

Pages 72–73 Fresno, California, 1901. Klein, L. W., S[an] F[rancisco], Britton & Rey, c1901. G4364.F8A3 1901 .K5

Pages 74–75 Bakersfield, Kern County, California, 1901. Photo-lith. Britton & Rey. Stone (N.J.) Company. San Francisco, c1901. G4364.B2A3 1901 .S7

Pages 76–77 Eureka, Humboldt County, California. Copyright by A. C. Noe & G. R. Georgeson. Photo-lith. Britton & Rey. Noe (A. C.) & G. R. Georgeson. [Eureka?] 1902. G4364.E9A3 1902 .N6

Pages 78–79 Berkeley. Green, Charles, draughtsman. Berkeley, Cal. [1909?] G4364.B5A3 1909 .G7

Pages 80–81 Greater Los Angeles: the wonder city of America, designed by K. M. Leuschner. copyrighted. Published by Metropolitan Surveys. Leuschner, K. M. G4364.L8 1932 .L4

Page 82 View of Sacramento City as it appeared during the great inundation in January 1850, drawn from nature by Geo. W. Casilear & Henry Bainbridge; lith. of Sarony, New York. PGA—Sarony (N.)—View of Sacramento City . . . (D size), Sarony, Napoleon, 1821–1896, lithographer. Prints and Photographs Division, [P&P] c1850.

Pages 86–87 Map of part of San Francisco, California, April 18, 1908: showing buildings constructed and buildings under construction during two years after fire of April 18, 1906. Punnett Brothers. San Francisco: California Promotion Committee, c1908. G4364.S5E73 1908 .P81

Pages 88–89 Earthquakes and faults in the San Francisco Bay Area (1970–2003) / by Benjamin M. Sleeter . . . [et al.]. Sleeter, Benjamin M. Menlo Park: U.S. Dept. of the Interior, U.S. Geological Survey; Denver, CO: For sale by U.S. Geological Survey, Information Services, 2004. G4362.S22C55 2003 .S5

Pages 90–91 Earthquake shaking potential for California, Spring 2003 / CSSC. California. Seismic Safety Commission. [Sacramento, Calif.]: CSSC, [2002] G4361.C55 2002 .C3

Pages 92–93 Topographic map of the Los Angeles aqueduct and adjacent territory / Water Department of the City of Los Angeles, Cal. Board of Commissioners. City of Los Angeles Board of Water and Power Commissioners. Los Angeles: City of Los Angeles, 1908. G4361.N44 1908 .C5

Pages 94–95 Southern California Wildfire Overview: 10/30/03, MAST, Mountain Area Safety Taskforce. [Sacramento, Calif.]: California Dept. of Forestry and Fire Protection, [2003] G4361.K5 2003 .M6

Page 96 The unique map of California. Johnstone, E. McD. [S.l., 1888] G4360 1888 .J6 TIL

Pages 100–101 John Steinbeck Map of America, (1986). Courtesy of Molly Maguire and Aaron Silverman

Pages 102–103 Raymond Chandler Mystery Map of Los Angeles. 1985 Literature/Film, *Language of the Land*, Courtesy of Molly Maguire and Aaron Silverman

Pages 104–105 Hollywood starland: official moviegraph of the land of stars, where they live, where they work and where they play, copyrighted by Don Boggs. Hollywood: 1937 G4364.L8:2H5A3.B6

Pages 106–107 Ragsdale's movie guide map: 1938 latest edition, copyrighted 1938 by Albert Ragsdale. Los Angeles: 1938. G4364.B6 1938 .R3

Pages 108–109 Map of Yosemite National Park, USGS, 1965. G4361.G52Y6 1965.T7

Pages 110–111 A hysterical map of Death Valley National Monument: and it's lookin' mighty low, drawn in broken English by Jolly Lindgren. Spokane, Wash.:Lindgren Brothers, 1948. G4362.D4E63 1948 .L5

Page 112 Plat of the exterior boundaries of Sequoia National Park, California: created by Act of Congress, approved October 1st, 1890, surveyed by Isaac N. Chapman, U.S. Deputy Surveyor . . . in July and August 1899 & May to September 1900. G4362.S42 1900 .C4

About the Authors

VINCENT VIRGA has earned critical acclaim for *Cartographia: Mapping Civilization* and co-authored *Eyes of the Nation: A Visual History of the United States* with the Library of Congress and Alan Brinkley. Among his other books are *The Eighties: Images of America* with a foreword by Richard Rhodes, *Eisenhower: A Centennial Life* with text by Michael Beschloss, and *The American Civil War: 365 Days* with Gary Gallagher and Margaret Wagner. He has been hailed as "America's foremost picture editor" for having researched, edited, and designed nearly 150 picture sections in books by authors including John Wayne, Jane Fonda, Arianna Huffington, Walter Cronkite, Hillary Clinton, and Bill Clinton. Virga edited *Forcing Nature: Trees in Los Angeles,* photographs by George Haas, for Vincent Virga Editions. He is the author of six novels, including *Gaywyck, Vadriel Vail,* and *A Comfortable Corner,* as well as publisher of ViVa Editions. He has a Web site through The Author's Guild at vincentvirga.com.

RAY JONES is a historian, author, and publishing consultant living in Pebble Beach, California. He has written more than thirty books on subjects ranging from dinosaurs to country stores but is probably best known for his lighthouse travel guides and histories. Published by The Globe Pequot Press in 2004, his award-winning *Lighthouse Encyclopedia* is widely regarded to be the best and most informative volume on the subject. He has also written a number of PBS companion books, including *Niagara Falls: An Intimate Portrait,* published in 2006, also by The Globe Pequot Press.